CORN MEETS MAIZE

CORN MEETS MAIZE

Food Movements and Markets in Mexico

Lauren E. Baker

ROWMAN & LITTLEFIELD PUBLISHERS, INC.

Lanham • Boulder • New York • Toronto • Plymouth, UK

Published by Rowman & Littlefield Publishers, Inc.
A wholly owned subsidiary of The Rowman & Littlefield Publishing Group, Inc.
4501 Forbes Boulevard, Suite 200, Lanham, Maryland 20706
www.rowman.com

10 Thornbury Road, Plymouth PL6 7PP, United Kingdom

British Library Cataloguing in Publication Information Available

Library of Congress Cataloging-in-Publication Data Available

ISBN: 978-1-4422-0651-9 (cloth)
ISBN: 978-1-4422-0653-3 (electronic)
ISBN: 978-0-8108-9509-6 (pbc : alk. paper)

∞™ The paper used in this publication meets the minimum requirements of American National Standard for Information Sciences—Permanence of Paper for Printed Library Materials, ANSI/NISO Z39.48-1992.

Printed in the United States of America

To Cameron Collyer, with love and gratitude

Contents

Preface and Acknowledgments

As I write these words, on the cusp of this book's publication, I am preparing for a workshop offered by Fulvio Gioanetto from Michoacán, Mexico, in Toronto. Urban and peri-urban farmers from the greater Toronto area will attend, and Fulvio will share his experience in agroecological farm management, using locally available plants and minerals to prepare biological pest, disease, and fertility applications. Fulvio spent the summer at Plan B Organic Farm in Ontario, working with Alvaro Venturelli and visiting many urban gardens and farms to speak with growers.

This book is about local food networks and their intersection with broader regional, national, and global social movements. It is about the circulation of plants, food, knowledge, and the practical work of building local food economies and transnational social movements. I focus on Mexico, where I had the opportunity to step away from my own work related to building sustainable food systems in Canada and learn from a different context, another culture, other issues and ways of organizing around food. The themes I explore in this book resonate deeply with my experience in Toronto, Ontario, and across Canada, and the intercultural global exchange continues to inform my current work at the Toronto Food Policy Council, as well as my participation in Sustain Ontario and Food Secure Canada at the provincial and national levels.

I would like to thank the Carrot Cache for supporting the ongoing agroecological exchange between Mexican and Canadian food activists and farmers. I am grateful to the faculty of environmental studies at York University for the Graduate Fellowship for Academic Distinction, which supported this research. Grants and awards from the Social Sciences and Humanities Research

Council of Canada, the Ontario Graduate Scholarships Program, and the International Development Research Centre's Ecosystems and Human Health Program provided funding for my fieldwork.

In 1995 I became involved with Deborah Barndt's Tomasita Project, and I feel honored to have worked with Deborah in various contexts ever since. Her encouragement, insight, and creativity inspired my research. Deborah's commitment to transnational collaboration, critical pedagogy, and social justice continues to inform and shape my work. There are many others to thank for stimulating discussion and debate—too many to name, but I can't help but list a few: Edit Antal, Kirsten Appendini, Gustavo Esteva, Debbie Field, Harriet Friedmann, Jane Hayes, Cathleen Kneen, Anan Lololi, Mary Lou Morgan, Wayne Roberts, Gabriel Torres, and my colleagues at the Centre for Studies in Food Security at Ryerson University.

My fieldwork would not have been possible without Amado Ramírez Leyva and Gabriela Fernández Orantes, who welcomed me to the Itanoní Tortillería. Their work has evolved into a new initiative "maize from Oaxaca's ancestral communities, a network of production, consumption and exchange." Jorge Aguirre Alonso and Armando Joffre Poceros introduced me to ANEC's Nuestro Maíz project. Fulvio Gioanetto and Aguilar Villaseñor Wenceslao hosted me in Michoacán. Thank you to my colleagues at El Colegio de México, Universidad Nacional Autónoma de México, and Universidad Autónoma Chapingo.

My family has provided immense support throughout the writing of this book. Samuel and Henry, my two exuberant boys, provided welcome diversion and laughter. Carole Baker nourished my family and this research with her love, interest, and insight. Jack Baker provided wonderful conversation and guidance. Eloise Graham offered her healing gifts and the delight of an expanded family. Gerry and Marie Collyer were a source of unwavering acceptance and encouragement. My sisters, Joanna and Amber, sustained me with their friendship.

Lauren Baker
Fall 2012

1

Why Food? Why Corn and Maize?

> Most people experience contemporary globalization as threatening local and microregional self-reliance, diminishing participatory democracy, and destroying diversity of cultures, while devouring the last remnants of natural resources and wilderness. In the rural areas, this perverse process is being made possible in all those spaces where individuals and communities are unable to resist the three main mechanisms of globalization: dependency, specialization, and centralization of power.
>
> Challenging these forces, a myriad of rural communities in countries as different as Finland, Japan, India, Australia, Peru, and Mexico have initiated, in the last decades, successful experiences of local organization and control over the global forces. . . . Diversity, self-sufficiency, grassroots democracy, equity, and decentralization of power are basic principles guiding the actions of these local movements. As a whole, these principles represent new paradigms in the construction of an alternative modernity; that is, the creation of a sustainable, ecologically inspired, post-neoliberal society.
>
> —Toledo 2001, 472

Local Movements, Global Window

MEXICANS ARE ORGANIZING "In Defense of Maize." This slogan has become the rallying cry for a food movement grounded in the agricultural biodiversity of maize, the country's staple food. The movement confronts global trends reshaping food and agriculture: neoliberal policy, corporate concentration, and dietary transformation.

PHOTO 1.1
Wonder tortillas made from *harina de trigo* are readily available in Mexican corner stores and supermarkets, reflecting agricultural and dietary transformation.

Every October, Mexico celebrates El Dia de la Raza, a national holiday in recognition of Mexican culture and traditions. In recent years, indigenous and civil society groups across Mexico used El Dia de la Raza to demand a ban on genetically modified (GM) corn. Monsanto has lobbied for years for the right to plant GM corn trials in parts of Mexico, something that activists and indigenous communities fear will contaminate native maize landraces. In the fall of 2009, the Mexican government approved trials of GM corn in designated areas, reigniting the movement In Defense of Maize.

The coalition In Defense of Maize gained strength in 2001 when scientists tested native maize landraces and found traces of GM corn had introgressed into, or contaminated, the gene flow. This shocking discovery gave rise to a social movement rooted in communities across Mexico but linked to global food sovereignty efforts. *Tortillerías* began to advertise that the maize they used was 100 percent maize mexicana, a protest against the cheap and poor quality maize imports from the United States. The statement "somos hijos de maize"—we are the children of corn—became a popular movement slogan.

In this book, corn is used as a symbol for the commoditization of food and the corporate control of food production, processing, and consumption. Maize, on the other hand, is used as a symbol to describe agricultural and food practices grounded in practical farming knowledge, culinary traditions, and local economic exchanges. These two symbols are used as a window into understanding key debates and tensions in the field of food studies: food security and food sovereignty, biodiversity and cultural diversity, culture and nature, the local and the global, markets and movements. The meeting of corn and maize offers new ways to understand these debates as fluid and mutually contested.

The struggle between maize and corn in Mexico is symbolic of how contemporary agricultural and dietary transformations are permeating local places. In the story of maize in Mexico we see the roots—structural and metaphorical—of the global food crisis; the ways that the obesity epidemic is related to the loss of culinary culture and skills; the poverty of a diet based on processed sugar, maize, and fat; the alienation of farmers from the land and practical skills of farming and seed saving. But it is also the story of hope. Communities and social movements are acknowledging these crises and reclaiming control of their food systems. In the process they are regenerating local ecologies, cultures, and economies, and contesting neoliberal trade policies (Pechlaner and Otero 2008).

These efforts—strategies developed by local communities in Mexico to conserve and protect maize agrodiversity—intersect with the strategies of food movements around the world that are forming new food supply and distribution networks and articulating the cultural and ecological importance of local food systems.

Victor Toledo, in the introductory quote, articulates the characteristics of the myriad local movements challenging globalization. The principles of diversity, self-sufficiency, grassroots democracy, equity, ecological integrity, and decentralization of power, Toledo states, represents a new paradigm from which a postmodern, post-neoliberal, ecological society is being constructed. Tim Lang and Michael Heasman describe the current context as a contest between the "Life Sciences Integrated paradigm" (characterized by specialized approaches to industrial agriculture and food practices) and the "Ecologically Integrated Paradigm" (characterized by ecosystem approaches to health) (Lang and Heasman 2004). Harriet Friedmann (2005a) suggests that we are witnessing the emergence of a new food regime. A myriad of forces are shaping the emerging food regime, and Friedmann argues that social movements play an important role as paradigms compete to succeed. Friedmann states, "A new regime seems to be emerging not from attempts to restore elements of the past, but from a range of cross-cutting alliances and issues linking food and agriculture to new issues. These include quality, safety, biological and cultural diversity, intellectual property, animal welfare, environmental pollution, energy use, and gender and racial inequalities" (249). The stories of corn and maize in this book tangibly demonstrate the competition between paradigms and the unfolding characteristics of the new food regime (for a summary of food regime formation, see Fairbairn 2010).

Three initiatives in defense of maize are used as examples of innovative models for organizing alternative food networks. Itanoní Tortillería, Nuestro Maíz, and the Michoacán Centre for Agribusiness are efforts to promote regional maize markets and protect campesino farmers from the incursion of cheap corn from the United States that has flooded the Mexican market since the North American Free Trade Agreement (NAFTA) was signed in 1994. Nuestro Maíz provides an example of how, in Mexico, campesino organizations are adapting to neoliberal policies and seeking regional, value-added markets for maize grown by their members. The Nuestro Maíz experience reveals the challenge of coupling rural development and agrodiversity conservation goals in the neoliberal context. The work of Itanoní Tortillería demonstrates the importance of education and skill building as part of local food movements. The Tortillería encourages eaters to appreciate the *taste* of maize diversity and works with farmers to improve and diversify their production practices. The Michoacán Centre for Agribusiness illustrates a number of activities governments can undertake to market local products more widely, both complementing and challenging organic export opportunities.

Beyond linking farmers to eaters, these tortilla supply chains are connected to transnational social movements, demonstrating how local food systems are permeated by global flows of knowledge, cuisines, seeds, markets, migration, and solidarity. The ways that "corn" meets "maize" in these initiatives illustrate the complexity of reclaiming food from its commodity status in the global context of financial turmoil, food crises, and climate change. As local food movements and markets firmly take root in communities across the globe, scholars, activists, and entrepreneurs are looking for new models and stories about the ways alternatives are being developed. More broadly the three initiatives represent different approaches to creating alternatives to the global industrial food system, and provide insights into conceptual challenges related to rebuilding local food economies. These conceptual challenges are the thorny issues emerging from the theory and practice of food system relocalization: the ways these initiatives intersect with broader policy processes; how they address issues of scale; the possibilities for (and limitations of) expanded notions of the economy; how to create organizational structures that embed food and farming in appropriate ecological, cultural, social, and economic systems; and how to take into consideration the agency of diverse actors, both human and nonhuman.

Corn Meets Maize is organized into six chapters. In the rest of this introduction, I discuss the key tensions explored through the two metaphors, corn and maize: food security and food sovereignty, biodiversity and biotechnology, culture and nature, and global and local. The initiatives are further examined. The text boxes found throughout the book allow me to link the story of maize and corn in Mexico to my work in the Canadian food movement, acknowledging the ways that my experience building sustainable food systems in Ontario and advocating for food and agriculture policy reform in Canada has been shaped by the work of others doing the same in Mexico and other places. Chapter 2, "Milpas, Markets, and Movements," offers a short history of maize and corn in Mexico, describing the dynamics of agrodiversity as both rooted in particular places and cultures, and reflective of "transworld" agriculture (Coleman 2003). The following three chapters delve into the initiatives. The three initiatives—Itanoní Tortillería, Nuestro Maíz, and the Michoacán Centre for Agribusiness—reveal different aspects of contemporary agricultural and dietary transformation in Mexico. They are examples of shortened food supply chains, linking farmers to eaters in their region and beyond. Beyond the linkages from farmers to eaters, all three initiatives have linkages and connections with broader social movements, civil society organizations, and researchers. They are connected to biodiversity conservation networks such as seed banking, agrodiversity education and agronomists, plant geneticists, and anti-GM activists. The initiatives illustrate the ways that agrodiversity and food networks are fluid and shaped by the cultural, ecological, institutional, and economic contexts in which they are embedded. Place-based, these initiatives are grounded in particular ecologies, cultures, and economies. And yet the projects are not place-bound. They are networked locally, regionally, translocally, and transnationally through the neoliberal policy context, the circulation of maize and corn via agriculture and food commerce, and social movements. The book's concluding chapter describes the practical and theoretical implications of the initiatives, demonstrating how local food networks are examples of biocultural agrifood relations, a concept that is explored throughout.

BOX 1.1
A View from the Roof: Urban Agriculture and My Own Translocal Roots

At a community food distribution warehouse in Toronto, I climbed up a ladder to the roof to see an urban agriculture experiment a friend was working on. Looking out over the vast concrete sea of empty rooftops, I was struck by the potential for the small area of lettuce that was planted on the roof I stood on. What if, I thought, what if we could transform all these roofs into productive spaces. How would that change the city? And so started an incredible urban agriculture adventure that led to a deep interest in local food initiatives.

My interest in maize and corn in Mexico emerged from this deep and long-standing involvement in food networks in North America. After working on food projects in Ontario for several years, I applied to start a master's of environmental studies at York University in 1995. As a graduate assistant for the Tomasita project, I accompanied Deborah Barndt to Mexico to unveil the journey of a tomato from Mexican field to Canadian table and the role of women workers along the trail.[1] While in Mexico I observed children working long days picking tomatoes beside their mothers, indigenous communities uprooted from their homes living in cardboard camps and working to pick the food we eat in the United States and Canada, and pesticide packaging strewn in roadside ditches beside the fields. This visceral experience of the global food system inspired me to work creating alternative food networks in Toronto upon graduation. I started this work by creating two small social enterprises, Annex Organics and Urban Harvest, based out of FoodShare's Field to Table warehouse, and eventually began working full time as FoodShare's urban agriculture program coordinator.

The urban agriculture work was grounding and tangibly part of the process of creating life. The simple act of placing a seed in soil, watering it, watching it develop true leaves and mature over the season to produce fruit. Watching nature's resilience every day at an urban warehouse surrounded by contaminated industrial brownfields. Over the five years I worked at the Field to Table warehouse these simple acts of planting shifted my conceptions of nature and its place in urban areas.

I became interested in food plant diversity. Choosing new varieties to grow each year involved spending hours every winter reading seed catalogs. Many of these endangered varieties had captivating stories, and reading the seed catalogs was like a trip into a botanical garden for unusual and rare food plants. I read the book *Lost Crops of the Incas: Little Known Plants of the Andes with Promise for Worldwide Cultivation* (National Research Council 1989) and experimented with some of the interesting plants—*oca*, *mace*, and *cape gooseberry*—on our rooftop garden. I was fascinated by the history of these plants and their interesting culinary uses and struck by the corresponding lack of diversity on our supermarket shelves.

The warehouse became an unlikely diversity garden. Volunteers and visitors to the warehouse brought seeds that were planted and grown out. These seeds reflected their favorite foods, their gardens, their agricultural and culinary history. The links between Toronto's ethnocultural communities, agrodiversity, and urban landscapes were revealed.

When I decided to pursue a doctoral degree, I had recently completed a participatory research project with eight community groups across the city of Toronto that examined the intersections between cultural diversity and biodiversity.[2] This project piqued my interest in these intersections, revealing there was much to be explored about the global biocultural diasporas in Toronto's community gardens.

Walking through the Leslie Spit allotment was like a tour through a hundred aesthetic sensibilities—plots full of roses carefully manicured and trellised next to a cacophony of wildflowers, beside a vegetable plot planted in careful rows. Most interesting to me were the unfamiliar plants in the vegetable plots of the Frances Beavis community garden for Chinese seniors. At one point in the project we brought local farmers to visit the garden to see the innovative production methods used to polycrop Asian greens and squashes. I was captivated by the gardens located in the inner suburbs of Toronto, landing places for new immigrant families seeking a new home in Canada but maintaining their agricultural and culinary roots through the community garden.

It was through these gardens and with these gardeners that I began to understand the interrelationships between biodiversity and cultural diversity as ontologically related. The diasporic reality of the gardeners encouraged me to think about how the gardens and gardeners were place-based, yet not place-bound due to their ongoing connections and histories in other places.

At the same time I was enjoying these urban gardening experiences, I was gaining a practical understanding about the politics of food. I was working with youth at risk, mentally ill, and low-income people in Toronto. These people understood hunger and could describe what it felt like to crave healthy, nutritious food, but only have the offerings from the food bank in the cupboard. FoodShare's work involved linking these eaters to farmers and innovating ways to offer fresh, healthy food to low-income Torontonians. The daily contrasts of planting heritage vegetables and packing food boxes helped me to understand how food plant diversity loss is connected to the industrial food system and health.

Over the duration of my work in community food security in Toronto I had become increasingly interested in the meaning of community food practices and wanted to delve into this theoretically. A priority for me was rekindling my connections to Latin America, generally, and Mexico, specifically. This desire was grounded by a deep commitment to north/south connections and translocal solidarity.

In 2001 and 2002, as I was beginning my doctoral degree, the maize debates in Mexico caught my attention. This was a local food issue that had galvanized international attention, involving disparate actors including scientists, policy makers, academics, and activists both locally and globally. Through different fora, links were being made between culture, ecology, and economy in new and interesting ways. During an exploratory trip to Mexico in the winter of 2004, I sought out ways that community groups were building local food networks in response to the issue of maize agrodiversity and to address the local repercussions of NAFTA.

These experiences and interests are the root of this book, which explores the interconnections between cultural diversity and biodiversity within globalized economic systems. My desire was to take a leave from the Canadian context and to seek a comparative perspective. I was convinced that "following" maize would allow me to get inside the global food system and explore the interconnections between biodiversity and cultural diversity in local food networks.

From my work on community food security in Ontario I understood that food was a valuable theoretical and pedagogical device. Many of the studies that inform my methodology follow food through the global food system, across countries and continents (Cook et al. 2006).

My experience working in local food networks—food initiatives working to connect local farmers to eaters, espousing the principles of trust, quality, equity, and environmental sustainability—was that, although reflecting a shorter food supply chain, they were equally embedded within global cultures, markets, and ecologies. In this research I set out to explore the "global nature" of local food networks to find what stories they can tell us about the global food system.

Notes

1. My own experience of this project was published in 1999: "A Different Tomato: Creating Vernacular Foodscapes," in *Women Working the NAFTA Food Chain: Women, Food and Globalization*, ed. D. Barndt, 249–260. Toronto: Second Story Press.

2. This project was documented in several publications. I wrote a report for FoodShare titled *Seeds of Our City: Case Studies of Eight Community Gardens in Toronto* (2002); an article with Jin Huh in *Alternatives Journal* (2003), "Rich Harvest"; and an article in *Geographical Review* (2004), "Tending Cultural Landscapes and Food Citizenship in Toronto's Community Gardens."

Food Movements and the Food Security/Food Sovereignty Debates

In 1974 the World Food Council was established by the Food and Agriculture Organization (FAO) to address a global food crisis marked by famines in Africa and pressure on the international grain market due to increasing global demand and rising grain prices. This historical moment, the discourse of that food crisis, and the intensification and broadening of the notion of food security at the time echoes resonantly. Today, in a similar fashion, the global food economy is straining under the demand for biofuels, growing consumption of grain-fed meat, and production and distribution systems that rely on high cost oil. The transnational food distribution system is not serving a large percentage of the global population with many people either over- or undernourished. Local agricultural economies and cultures are threatened by industrial agricultural practices. New actors have emerged to address these concerns. The examples in *Corn Meets Maize* are situated in the shifting food security–food sovereignty debates of the twenty-first century.

The concept of food security has changed dramatically over the past fifty years, from a focus on nutritional requirements of individuals and communities and the physical availability of food at a macro level, to the recognition that food and agriculture policies interact with complex socioeconomic factors at various scales to impact food security (Maxwell 1996). The food security discourse now recognizes that livelihood strategies, social networks, and the political, economic, and environmental context are key factors that determine food security, defined by the FAO as "when all people . . . have access to sufficient, safe and nutritious food to meet their dietary needs and food preferences for an active and healthy life" (FAO 2003).

The definition continues to be critiqued and expanded. The Centre for Studies in Food Security at Ryerson University in Toronto, Canada, defines the key characteristics of food security as: availability (sufficient supply); accessibility (effective distribution); acceptability (culturally acceptable/nutritionally adequate); appropriateness (ecologically sustainable, safe); and agency (enables action) (Centre for Studies in Food Security n.d.). This definition inserts cultural and environmental factors to augment the focus on participation and process as key components of food security (See also Koc et al. 1999 and Maxwell and Smith 1992).

The limitation of the food security discourse is its focus on food security as a material condition related to deficiencies in production and distribution. Many responses to food insecurity have reflected this focus and encompass redistributing food in emergency contexts, developing charitable distribution systems, and using food as a community development tool to increase people's ability to purchase and prepare healthy food (see, for example, Koc et al. 1999 for a review of these approaches).

Emerging with a strong challenge to the food security discourse has been the political platform of food sovereignty. Coined by the international peasant coalition Via Campesina in 1996, food sovereignty promotes alternatives to neoliberal development. At the Forum for Food Sovereignty in Mali in 2007, delegates adopted the following declaration that describes the principles of the concept and movement:

> Food sovereignty is the right of peoples to healthy and culturally appropriate food produced through ecologically sound and sustainable methods, and their right to define their own food and agriculture systems. It puts those who produce, distribute and consume food at the heart of food systems and policies rather than the demands of markets and corporations. It defends the interests and inclusion of the next generation. It offers a strategy to resist and dismantle the current corporate trade and food regime, and directions for food, farming, pastoral and fisheries systems determined by local farmers. Food sovereignty prioritises local and national economies and markets and empowers peasant and family farmer-driven agriculture, artisanal fishing, pastoralist-led grazing, and food production, distribution and consumption based on environmental, social and economic sustainability. Food sovereignty promotes transparent trade that guarantees just income to all peoples and the rights of eaters to control their food and nutrition. It ensures that the rights to use and manage our lands, territories, waters, seeds, livestock and biodiversity are in the hands of those of

> us who produce food. Food sovereignty implies new social relations free of oppression and inequality between men and women, peoples, racial groups, social classes and generations.

The food sovereignty movement moves beyond the food security discourse to directly address the structural issues related to land tenure, macroeconomic policies, human rights, and local control of decision making and resources. Many of the demands quoted above focus on the linkages between controlling the means of social reproduction and global, neoliberal economic policy. If food security describes and prescribes solutions to a material condition, food sovereignty understands these intersecting issues as a historical economic, ecological, and cultural process. The intimate interrelationships between people, the land, cultural practices, and local economies are explicitly described (Fairbairn 2010).

Philip McMichael offers a useful analysis (2008) of the food sovereignty movement, contrasting food sovereignty and food security. Food security is a manifestation of the neoliberal project and is based on "installing a 'self-regulating market' on a world scale [that] *encloses* questions of social reproduction within a legitimating rhetoric of 'feeding the world'" (McMichael 2008, 216). Three key shifts are articulated by McMichael in his work on food security and food sovereignty. First, is the shift from a focus on production to a focus on social reproduction. This is an epistemological shift from viewing food beyond its material value to embodying social, cultural, and ecological values (218). Second, McMichael demonstrates how the food sovereignty movement has revealed the state's complicity in the neoliberal project, a "capital/state nexus," through policies at various levels from the global to the local, and aims to transform the state system (212). Third, the food sovereignty discourse represents an ontological shift. McMichael explains,

> The ontology of the food sovereignty movement critiques the reductionism and false promises of neoliberalism, positing a practice and a future beyond the liberal development subject, and the science of profit. This emerging ontology is grounded in a process of revaluing agriculture, rurality and food as essential to general social and ecological sustainability, beginning with a recharged peasantry. (2008, 213)

The agrifood movements that reflect and shape the food security/food sovereignty debates address the social, economic, and environmental costs of the industrial food system and global trade regimes, as well as regenerate local food systems (Allen et al. 2003). Varied in their focus, agrifood movements encompass efforts to challenge corporate concentration, improve food safety, link eaters to farmers, encourage local food consumption and greater food self-sufficiency, address hunger and poverty, and promote sustainable agriculture (Pretty 2002). These movements have also been referred to as food networks or food webs (Hinrichs 2003), terms that acknowledge the complex and nonlinear household, community, regional, and institutional relationships involved in getting food from the field to the table. Although local food movements are often rooted in place—historically, culturally, and geographically—they are also connected transnationally through the modern diaspora and social movement networks (Friedmann and McNair 2008).

Local food movements create economic possibilities that challenge global capitalist food systems characterized by environmental and social exploitation. J. K. Gibson-Graham (2006) theorizes these economic possibilities by challenging us to move away from the dichotomies of global/local and conventional/alternative. Acknowledging that local food movements are embedded in (even as they challenge) a global neoliberal context, *Corn Meets Maize* reveals some of the overlapping and competing sociocultural, political, and economic relations shaping local food networks in Mexico.

Agrifood movements have been described and theorized from a number of disciplinary perspectives. For example, Allen (2004) examines sustainable agriculture and community food security movements in the United States through the lens of social justice movements, analyzing their discourses and practices. Goodman describes how local food networks represent "a movement from the 'industrial world,' with its heavily standardized quality conventions and logic of mass commodity production, to the 'domestic world,' where quality conventions embedded in trust, tradition and place support more differentiated, localized and 'ecological' products and forms of economic organization" (Goodman 2003, 1). McAfee (2006) describes how food sovereignty movements are bringing together disparate rural and farmers movements with international NGOs (nongovernmental organizations) to address global trade injustices. Statements from the Via Campesina and international food sovereignty meetings illustrate how sociocultural, economic, and ecological issues are intricately connected for movement activists and farmers (FoodFirst 2003).

Research on the proliferation of food movements has focused on disparate issues such as scale, the challenges of bridging social justice and environmental issues, and the limitations of local projects (Allen et al. 2003; Hinrichs 2003; Johnston 2003; Johnston and Baker 2005). In this literature food movements are articulated as political spaces that engage people in democratic practices that occur as part of everyday life and simultaneously impact policy at various levels.[1] These democratic practices are part of creating "food citizens" who are not only eaters, but who are also engaged in their communities and have an "intimate" connection to the food they eat (Winson 1994; Welsh and MacRae 1998; DeLind 2002; Hassanein 2003; Wekerle 2004).

The failure of corporate, industrial agriculture to feed the world manifests itself differently in the north and south. This is reflected in the different approaches of agrifood movements in these contexts. In the north, these movements tended to focus on developing more intimate linkages among farmers and eaters, organic and ecological farming practices, and community food projects (see Allen 2004). Many of these elements can be found in the projects of southern agrifood movements, but the focus in this context has tended to be on developing fair trade networks, advocating for global policies that support small-scale peasant food production practices, and land reform (Holt-Giménez 2006; Goodman 2003).

In this section I have briefly outlined the food security/food sovereignty debates. I describe food security as reflecting a material condition with solutions aimed at addressing inadequate production and distribution systems. Food sovereignty, on the other hand, is a response to a historical process requiring political and structural transformation. These debates reveal how food movements can address either or both of the material condition of food security and the transformational politics of food sovereignty.

Biodiversity and Biotechnology: Diversity Debates

Intersecting and overlapping with the food security–food sovereignty debates are the biodiversity-biotechnology debates. The intensification of industrialized farming and global food distribution practices since the 1950s has threatened food crop diversity.[2] Cary Fowler and Pat Mooney report a staggering loss of food plant genetic resources since 1903—only 3 percent of vegetable varieties survived between 1903 and 1982 (Fowler and Mooney 1996). Modern, improved, and hybrid varieties displaced old varieties and landraces, a trend that was compounded by Green Revolution technologies.[3]

This loss of agrodiversity is worrisome for a number of reasons. Agricultural productivity is reliant on varietal diversity to address new and ever-changing crop breeding problems such as drought resistance and disease (Brush 2004). Small-scale farmers are equally reliant on the local ecological adaptability of this varietal diversity. Sociocultural practices—culinary, spiritual, and community—depend on the availability of different varieties for multiple purposes and reflect local knowledge about agriculture and ecology. The issue of agrodiversity loss is being addressed at the international, national, and local level through a number of conservation efforts.[4]

This book examines three initiatives working to conserve maize agrodiversity in Mexico by directly marketing maize landraces to eaters through the creation of local markets. The initiatives illustrate the possibilities for and challenges of maize agrodiversity conservation through local food networks, as well as the interconnections between cultural diversity and biodiversity.

Compounding the general trend of agrodiversity loss is the specific threat of GM seeds and agricultural biotechnology. The impact of biotechnology on agricultural biodiversity is being debated in a number of bio-political arenas (Royal Society et al. 2000). Proponents of biotechnology state that GM varieties will increase global food security (Elderidge 2003). Opponents state that the ecological and health ramifications of biotechnology are unknown and that the benefits of new technologies do not reach economically marginalized communities nor the farmers they are purported to serve (Altieri 2000).

In 2001, the scientific journal *Nature* published a paper by David Quist and Ignacio Chapela reporting the discovery of transgene flow from GM corn to maize landraces in indigenous communities' fields in Oaxaca, Mexico (Cummings 2002).[5] This discovery has repercussions for local and global food security and the resilience of agricultural systems (Fowler and Mooney 1996). The discovery of this introgression intensified the work of translocal networks addressing maize agrodiversity and the ownership of genetic resources.

Mexico has been identified as a "centre of genetic diversity" for food crops (Brush 2004). The sociocultural importance of maize in Mexico, and the agroecological importance of maize diversity for global agriculture, has meant that agronomists, geneticists, crop improvement specialists, ethno-botanists, farmers, activists, and rural development workers have been working to document, categorize, conserve, and promote maize agrobiodiversity in Mexico since the early 1900s (Paczka 2003). The discovery of transgene introgression to maize landraces politicized and shed light on the importance of these conservation efforts, according to Flavio Aragon Cuevas, a maize agrodiversity specialist at INIFAP (the National Institute for Forestry, Agricultural and Livestock Research) (pers. comm., February 8, 2006). New actors became involved in promoting and defending maize agrodiversity in Mexico.

In this book I focus on contemporary maize debates in Mexico, but it is important to recognize that maize, historically, has been at the center of Mexican social and cultural movements (see Pilcher 2005). As Gerard Verschoor (2007) states, what has changed is how the issues are framed. I am interested in how the GM corn issue galvanized new ways of articulating the sociocultural, agroecological, and political-economic importance of maize in Mexico. Conflicts between biotechnology and biodiversity are at the heart of these debates, and the actors defining the claims have an interest in intellectual property, profits, and market control, on the one hand, and biosecurity, health, food security, agricultural biodiversity, sustainable rural livelihoods, indigenous rights, and cultural sovereignty, on the other hand. These biodiversity-biotechnology networks figure centrally in the "neoliberal corn regimes" (Fitting 2006) in the ways they challenge and reflect neoliberal policies at local, national, and international scales.

Sarah Whatmore in her book *Hybrid Geographies* (2002) illustrates how biodiversity-biotechnology conflicts are based on differences in the definition of diversity, using the FAO and Convention on Biological Diversity (CBD) as examples.

> Where the FAO version articulates diversity as a heterogeneous achievement in which human being and doing is enmeshed through long and situated association in the spatial and corporeal fabric of botanical becomings, the CBD account casts it in wholly biological terms, the outcome of an evolutionary process divested of human presence. The one conjures a world that is hybrid "all the way down," enfolding humanity in its ceaseless commotion time out of mind. The other conjures a world until recently unmarked by the (invariably negative) "impacts" of human society, only countenancing hybridity as a technological accomplishment associated with the advent of "genetic resources." (Whatmore 2002, 92)

In the case of maize, the mutually constituting association between plants and people over time clashes with static definitions of maize agrodiversity.

The discovery of transgenic corn in Oaxaca, Mexico, was a catalyst for both farmer groups and civil society organizations to organize local, national, and international campaigns that address a number of interrelated political, cultural, and ecological issues (Cummings 2002). These campaigns reveal the impact of neoliberal policies in their analysis of the global economic context; support farmers to maintain, preserve, and manage biodiversity; and advocate for policy change through their involvement in national and international policy processes (see, for example, Bartra 2005; Esteva and Marielle 2003). The intersecting issues of biotechnology and agricultural biodiversity are an example of how contemporary agricultural transformation impacts food security and food sovereignty. This book examines how civil society organizations, farmer groups, and government are organizing to respond to agricultural biodiversity and biotechnology conflicts (Altieri 2000; Pretty 2002). This is central to understanding how emerging technologies affect the most marginalized groups in society, and how these groups are organizing to defend their livelihoods and biocultural food practices.

The correlation between cultural diversity and agricultural biodiversity has not been well documented (Brush 2004),[6] though it has been suggested in several ethnographic studies. For example, Nazarea (1998) demonstrates a relationship between the conservation of genetic resources and the conservation of culture in indigenous sweet potato production in the Philippines, and Roberto González (2001) highlights a link between agricultural and cultural practices in the local farming systems of Zapotec farmers in Oaxaca, Mexico. Indigenous farmers in Mexico, in particular women, play a central role in the on-farm (in situ) preservation of biodiversity (through the cultivation of milpas—plots for household and community consumption) and in community food security (through their role in family food procurement and preparation) (R. González 2001). A study by the Commission for Environmental Cooperation titled *Maize and Biodiversity: The Effects of Transgenic Corn in Mexico* (2004) identifies a number of sociocultural concerns to transgenic corn, including the impacts of agricultural and dietary transformation and the deep cultural importance of maize in Mexico.

Similarly, Kathleen McAfee (2004) describes how blanket technological solutions have not historically addressed complex socioeconomic and ecological issues. McAfee calls for "more place-specific, multifaceted, and farmer-centred approaches to agricultural productivity and sustainability" and calls on us to acknowledge "the intricate relationships of food systems to local cultural practices and institutions and the intimate connections of both to ecosystems and biodiversity" (82).

The agroecology and ecosystem health literature also recognizes the interactions between agroecosystems, biodiversity, food security, and rural community development (Waltner-Toews 2004; Dalgaard, Halberg, and Fenger 2002; Forget and Lebel 2001; Allen et al. 1991; Altieri and Hecht 1990). The complexity of the relationships between sociocultural, physical, environmental, and economic factors within an ecosystem is articulated as "nested hierarchies" of interlocking elements (Forget and Lebel 2001, S12). "Ecohealth" research traces the connections between health, environment, and development, illustrating the role social and cultural practices play in determining health at the broadest level (Forget and Lebel 2001). As noted by Richard Levins, agroecology pioneer, at an international workshop called "Food Sovereignty, Conservation, and Social Movements for Sustainable Agriculture in the Americas," held at Yale University in 2004, agroecology is increasingly recognized as a social as much as a technological project (McAfee 2006, 9). Deborah Barndt concludes her study of women workers in the North American food chain by making the link between food, agriculture, and health and stating that "nurturing a more holistic conception of health—as well-being that is at the same time physical and spiritual, individual and collective, human and non-human—is a starting point. But it also requires action, collective and coalitional efforts that move beyond concern for our own bodies, to a commitment to the sustenance of the collective body, of the earth's body" (Barndt 2002, 226–227).

Despite this research, the social and political networks that enable agro-ecological practices and ensure food security have seldom been examined explicitly (Pretty 2002; McAfee 2004). This book expands the links between culture and biodiversity to include the agrifood movements that are mobilizing around the issue of sustainable agroecosystems, health, and food sovereignty. Local farmers and eaters in Mexico are organizing in defense of maize—articulating the importance of their sociocultural and agroecological practices—in response to current biotechnology and biodiversity debates.

The spectrum of agrodiversity efforts in Mexico include seed collection, documentation, and banking, as well as political action regarding GM corn and the importance of Mexican agrodiversity. Projects that focus on the use of maize landraces and on generating sustainable livelihoods from farming have proliferated, as have initiatives that focus on consumer education and appreciation of agrodiversity. Local markets for maize landraces and other products, including coffee, amaranth, greens, chili peppers, and squash, are being developed. This activity, as well as a number of other campaigns in defense of maize, illustrates the plural politics of maize in Mexico, with strategies that encompass political organizing, coalition building and advocacy, changing consumption practices, and developing alternative commercial outlets.[7]

These freshly politicized maize agrodiversity conservation efforts are embedded in national and global economic and trade policies and respond to and reflect the ongoing transformation of the Mexican countryside. Rural transformation has taken place in a context where international and national agriculture and trade policies and rural development strategies are entangled and inseparable from their impact on rural and indigenous communities. The political economy and ecology literature (see, for example, McAfee 2006, 2004; Barkin 2003; Appendini 2001) thoroughly documents the environmental and social effects of economic liberalization on maize production in Mexico. In brief, since 1988 maize farmers in Mexico have experienced the dismantling of a national system of government supports and price stabilization mechanisms. The effects of this change in rural Mexico were intensified in 1994 with the signing of NAFTA and the opening of Mexican markets to imports of corn from the United States. These trends reflect the International Monetary Fund's structural adjustment policies and global efforts to liberalize trade, promoting export-oriented agro-economies to pay off the Mexican debt and giving comparative advantage to the larger, more subsidized agricultural sector in the United States. Chapter 2 expands upon the dynamics of maize diversity in Mexico, and this book seeks to understand these intricate relationships and connections in the context of local food networks.

Contentious Kernels: Culture and Nature, Global and Local, Markets and Social Movements Intertwined

In *The Omnivore's Dilemma*, Michael Pollan describes how North Americans in the twenty-first century are "maize walking" (2006, 19), a bodily expression of the dominance of corn derived products in the food we eat. Maize is no longer primarily consumed by North Americans in its raw, relatively unprocessed form as a grain, but instead is eaten in huge quantities in the form of high fructose corn syrup found in processed foods and as meat (poultry, cattle, pigs) raised almost exclusively on a corn-based diet. Because corn photosynthesizes carbon differently from most plants, carbon compounds derived from corn can be quantified in the human body,[8] revealing the uniform North American diet. Surprisingly, comparative analysis of these carbon compounds in American and Mexican hair or flesh shows a much higher ratio of carbon isotopes derived from corn in the American population (Pollan 2006, 22–23).

Although physiologically "corn walking," Americans generally do not think of themselves as having a corn-based diet, and certainly do not articulate deep cultural roots related to corn, the way Mexicans do with maize. The dominance of corn in the U.S. and Canadian diet (and beyond) is representative of an agroindustrial relationship based on specific dietary and agricultural practices that are characterized by resource-intensive, monocrop farming systems, centralized food manufacturing and distribution processes, and processed foods. Moving south across the U.S.-Mexican border, however, the relationship between corn, maize, and culture shifts from a primarily agroindustrial relationship to a deeply rooted, sociocultural and agroecological relationship.

According to the Popol Vuh, the Mayan creation myth, Mesoamericans are the original people of maize. The agricultural, cultural, and culinary anchor of Mexican society, maize is a common theme in art, policy, academic discourse, activism, and popular conversation. "If you talk about maize in Mexico people are interested. The theme of maize is always present in our conversations," states Gustavo Esteva, a Mexican activist (pers. comm., February 17, 2006).

Maize was domesticated in Mexico from its wild relative, *teosinte*, about nine thousand years ago (Bellon and Berthaud 2006, 4). Mexico is commonly known as the center of origin for maize domestification and diversity and fifty-nine landraces have been documented (Paczka 2003, 133). Maize diversity in Mexico is a result of complex biological and sociocultural interactions among local environments, farmers, and maize plants, and illustrates the interconnections between biodiversity and cultural diversity. Maize diversity today is conserved and perpetuated through farmer management practices in ecologically diverse regions of Mexico. Also contributing to agrodiversity conservation in Mexico are household and culinary uses of maize diversity, as well as national and international ex situ and in situ conservation strategies. Maize diversity in Mexico has been shown to be extremely important for global agriculture through its ability to supply germplasm or genetic resources that contribute to modern plant breeding and the development of hybrid varieties (Fowler and Mooney 1996).

Mexico is known as a center of origin for agricultural biodiversity, a concept that links ethnocultural diversity to the diversity of cultivated plants. Nikolai Ivanovich Vavilov (1887–1943), a biologist, geneticist, and plant collector from the Soviet Union, coined the phrase *center of origin* and developed a theory of food crop diversity that challenged the conventional thinking of his day. Previous to Vavilov, agriculture was thought to have developed along the Tigris and Euphrates rivers. Vavilov was able to show that agricultural biodiversity was concentrated in the latitudinal (Europe and Asia) and longitudinal (Americas) mountain ranges where varied topographical and climactic conditions both nurtured and sheltered diversity (Fowler and Mooney 1996, 31–32). Vavilov linked the ecological diversity and cultural diversity of these regions to the diversity of agricultural crops. His research led him to pinpoint eight centers of origin around the world. Further research has since expanded Vavilov's theory, but the centers of origin concept is still widely used and accepted.

Mesoamerica, in addition to being the center of origin for maize, is the origin of fifty edible fruits, herbs, vegetables, and grains. Vavilov's research, along with contributions from other biologists, geneticists, anthropologists, and archeologists, demonstrates the dense and complex web of social and ecological relations that support and nurture food crop agrodiversity. Central to this web of relations are the knowledge, preferences, and practices of Mexican farmers who develop and maintain agrodiversity. Fowler and Mooney, in *Shattering: Food, Politics, and the Loss of Genetic Diversity* (1996), extend this web of relations beyond the farmer and fields to connect food plant diversity with the political factors that shape global food security.

As a cultural artifact, maize is entwined with farming and culinary practices in Mesoamerica and beyond, and is a central part of the story of agricultural and dietary transformations globally. Following maize throughout history provides a lens through which to understand the processes of colonialism, development, and globalization. In the same way that these processes overlap, they are contingent and partial. Along the way, maize has intervened and shaped technologies, actors, policy processes, social relations, and agricultural and dietary practices.

In what follows I offer three vignettes that illustrate how the story of maize sheds light on the dynamic interplay between nature and culture, the global and local, and markets and social movements.

1. *Maize as natural/cultural artifact.* Unlike other cereal crops, maize cannot be traced directly to a wild relative. *Teocinte,* a wild and weedy grain found in Mesoamerica, is thought to be maize's closest morphological kin, but is so unlike maize as we know it that geneticists have not been able to demonstrate the evolutionary sequences resulting in a close relative to the maize we are familiar with today (Warman 2003).

Teosinte looks nothing like maize, has little nutritional value, and is not a practical food source. In contrast, the wild relatives of the other cereal crops are edible and highly nutritious. More importantly, these wild relatives have non-shattering[9] variants, selected by farmers over time during the process of domestification. There is no known non-shattering variant of *teosinte.*

Maize is different from the rest of the cereal crops in other aspects as well. Maize is the only cereal crop that if left unattended in the field, could not propagate itself. The tough husk surrounding the kernels make it impossible for maize to germinate. Even if the husk was removed, the kernels are so close together that they crowd each other out during germination. Botanists have called maize a "hopeless monster" (Fussell 1999) because of its inability to propagate itself without the help of a farmer. Some have gone as far as saying that maize was not domesticated, but created or invented by humans—"the outcome of some bold act of conscious biological manipulation" (Mann 2005, 218).

The intimate and untraceable connection between maize and farmers points toward the inseparability of nature and culture, how they are deeply entwined with one another.

2. *Global/local agricultural and dietary transformation.* At the time of colonial contact, maize was the center of Mesoamerican cultures, economies, and cosmologies. Maize spread relatively slowly from its center of origin in Mesoamerica longitudinally throughout the Americas due to the great variation in climates and soils. After colonization, maize was exported from the Americas to Europe and beyond, and because of the range of growing conditions swiftly adapted latitudinally to similar climatic zones (Warman 2003).

In the Americas, maize's dietary importance is inextricable from culinary practices that transform its nutritional content and make nutrients more readily accessible to the human digestive system. When maize was exported to Africa and Europe, with it spread two human health conditions—protein deficiency and pellagra (niacin deficiency) (Warman 2003). The exportation of maize without its complementary agroecological system and culinary traditions resulted in nutritional deficiencies never seen in the Americas. The milpa, the agroecological system in which maize is grown, provides Mesoamerican diets with complementary proteins, through beans and squash, to those in maize. The practice of soaking maize with an alkali (*cal* as it is known in Mexico—slaked lime from limestone) to remove the tough outer layer of the kernel and release amino acids such as niacin, ensured that maize, along with the other crops in the milpa, provided a nutritionally complete diet. In Europe pellagra became associated with maize when famine struck and peasants were forced to eat only ground maize (Warman 2003). In this context, maize became a symbol of poverty.

Maize is also a symbol of how capital penetrates and shapes agricultural processes. Maize is a highly important cereal crop globally due to the high yields that can be achieved per hectare. Arturo Warman, an anthropologist and former minister of agrarian reform in Mexico, offers one of the most complete histories of maize in his book *Maize and Capitalism: How a Botanical Bastard Grew to Global Dominance* (2003). Warman describes how maize became the staple food imposed by colonial and imperial powers during early agricultural transitions, fueling the slave trade and the colonial plantations throughout the nineteenth century. Tracing maize's movement from Mesoamerica to Europe, Africa, and China, Warman illustrates its role

> as a conqueror of wide-open spaces in China or the United States; of the seasonal agricultural calendar in southern Europe; of the organization and intensity of labor in Africa; as sustenance for workers in profitable, dynamic, and new enterprises; as a way to stave off hunger in the past and, I hope, in the future; and, lastly, as a symptom and stigma of poverty. (Warman 2003, 233)

Easily adaptable to diverse climatic and agricultural conditions, maize is an example of a colonial food import that was integrated into local diets as they started to reflect the global economy.

Along with the micro adaptations and transformations of maize in the fields and kitchens of the Americas, Europe, Africa, and Asia, more macro agricultural and dietary transformations were underway. Maize was central to transforming the global diet from one based primarily on plants to a meat-based diet as it became fodder and feed for livestock. Broad trends in agriculture have seen polycultural farming systems that included animal husbandry replaced by monoculture agriculture and confined animal feedlots dependent on grain, especially corn.

Easily processed into a number of agroindustrial products, corn continues to fuel dietary transformations. High fructose corn syrup found in processed food and soft drinks has been linked to the obesity epidemic. Corn is currently a key actor in the food versus fuel debate, shaping the discourse of the global food crisis and energy policies globally. It appears, from these historical and current accounts, that maize has been remarkably successful at enrolling the agricultural industry, farmers, scientists, policy makers, eaters, and entrepreneurs in its process of integrating into local and global food economies and cultures across the globe.

3. *Maize galvanizing markets and movements.* Maize did not gain prominence as a crop suited to industrial agriculture until after World War II when hybrids were created to produce uniform ears at a uniform height and harvesting combines were adapted for corn (Fussell 1999). At that time (post–World War II), maize was at the center of the Green Revolution as scientists, supported by the Rockefeller Foundation, began their inventory of maize varieties resulting in the current system of classification of maize by races (Fowler and Mooney 1996). The Green Revolution resulted in a shift in agricultural practices worldwide from polycrops to monocrops, from seed selecting and saving on the farm to purchasing patented hybrids, from closed farming systems to dependence on petrochemical-based fertilizers, herbicides, and pesticides. Most importantly, the Green Revolution signaled a shift from control of genetic diversity in the hands of farmers, to the control of genetic resources in corporate laboratories. In this sense, the biotechnology industry is an extension of the Green Revolution. Corn was one of the first GM crops and covering over 90 million acres is the most widely planted in the world (Food Democracy 2007).

The corn industrial complex—including livestock, high fructose corn syrup, and biofuels—is heavily subsidized in the United States and Mexico. The United States subsidizes corn farmers $122 per hectare, while Mexico pays farmers $53 per hectare. Eighty percent of subsidies in the United States go to 20 percent of corn growers, primarily agribusiness (Crone 2007). The United States subsidies result in an overproduction of low-cost corn, which gets dumped on the world grain market, hurting farmers in other places who cannot cover the costs of their production to compete with the deflated market price.

Corn has been at the center of many recent conflicts. In 2002 Zambia turned away thousands of tons of GM corn donated by the United States as part of a food aid package. The government was concerned that this corn would be planted and that the GM crop would threaten Zambia's trade with the European Union. In 2007 European Union environmental officials determined that GM corn could cause environmental damage and upheld a ban on cultivating GM corn. Furthermore, the European Union regulations require strict labeling and traceability of food containing GM ingredients. Also in 2007, the rising cost of corn on the global market due to a rising interest in biofuels caused tortilla prices to rise in Mexico and resulted in massive protests on the streets of Mexico City. Pechlaner and Otero (2008) describe how Mexico has been deeply impacted by neoliberal trade policy, combining their analysis of the social impacts of NAFTA with an examination of the adoption of and resistance to agricultural biotechnologies. They argue that resistance to both the neoliberal paradigm and biotechnology has, and is, shaping the neoliberal food regime (Pechlaner and Otero 2008, 12–17; see also Friedmann 2005a). These examples illustrate how maize galvanizes markets and movements and continues to shape global and local economies, agricultural technologies and practices, trade agreements, international aid policy, and scientific debates.

Corn Meets Maize: The Stories

The work of Itanoní Tortillería, Nuestro Maíz, and the Michoacán Centre for Agribusiness represent three distinct approaches to agrodiversity conservation, food sovereignty, and local and regional food networks in Mexico. The three examples illustrate how the material and symbolic meanings of maize shift and adapt to contemporary social, political, and economic manifestations. Grassroots responses to the "maize debates" and "neoliberal corn regime" in Mexico, the projects are working to create local networks to produce, distribute, and sell maize, tortillas, and other local food.

Mexico is the third largest nation in Latin America, after Brazil and Argentina. Topographically, the country is defined by four major mountain ranges: the Sierra Madre Occidental in the west, the Sierra Madre Oriental in the east, the Cordillera Neovolcánica in the center, and the Sierra Madre del Sur in the south. The coastal areas of Mexico, including the Yucatán Peninsula, are characterized by tropical or dry coastal lowlands. There is great climatic variation in Mexico because of the north-south extension and dramatic variations in altitude. This ecological heterogeneity and "rich mosaic of environmental and microenvironmental conditions" (Toledo 2001, 473) contribute to the incredible biological and cultural diversity found in the country.

The country's biocultural wealth includes a large number of species of flora and fauna, as well as fifty-four groups of indigenous peoples speaking 240 languages (Toledo 2001, 474). Three million peasant productive units, organized as *ejidos* and *comunidades* (although currently being dismantled under *ejidal* reform), also contribute to Mexico's biocultural diversity, allowing for small landholdings across topographically diverse landscapes. As Toledo writes,

> The profound relationship that has existed since time immemorial between nature and culture is a feature that can be witnessed in contemporary Mexico. Each species of plant, group of animals, and type of soil and landscape nearly

always has a corresponding linguistic expression, a category of knowledge, a practical use, a religious meaning, or an individual or collective vitality, for the indigenous inhabitants of the country. (2001, 474)

The three initiatives, Itanoní Tortillería, Nuestro Maíz, and the Michoacán Centre for Agribusiness, are based in Oaxaca, Puebla, and Michoacán states, respectively. Oaxaca has the second largest concentration of indigenous groups in Mexico and ranks first in terms of linguistic diversity with peoples speaking 104 languages (Toledo 2001, 476). Two-thirds of Oaxaca's land is managed by 1,406 small productive units.

Puebla is also bioculturally rich and topographically diverse, and has a population that is over 50 percent indigenous. The area under agricultural production is much smaller than Oaxaca (0.334 million hectares versus 8.048 million hectares in Oaxaca), and the number of productive units is correspondingly smaller (153 units) (Toledo 2001, 477).

Michoacán is most famous as the southern home for millions of monarch butterflies who migrate to the high altitude fir forests in the state. These butterflies have become a powerful symbol of the mass migration, since NAFTA, of Mexicans to the United States and Canada to work in the agricultural and food sector. Michoacán is the largest producer of avocados in Mexico, and local farmers have complied with international quality and organic standards to build export markets for fruit and vegetables.

The three initiatives are situated within these bioculturally diverse states and are examples of local food networks that link maize farmers to eaters. They illustrate the diverse ways in which new actors are responding to issues of maize agrodiversity and food sovereignty through the creation of local food networks in Mexico. The projects synergistically complement a wide spectrum of agrodiversity and rural development efforts in Mexico, but also illustrate the challenges faced by local food networks to simultaneously address social (rural development) and environmental (agrodiversity) objectives in the Mexican context, reflecting the challenges faced more generally by local food networks when attempting to address both social justice and environmental sustainability through their activities (see DuPuis and Goodman 2005; Johnston and Baker 2005; Allen 2004).

Itanoní Tortillería is a family-run restaurant that uses maize landraces grown locally by campesino farmers in the dishes served. The focus of their work is twofold—to support regional farmers to continue to grow maize landraces and to educate urban eaters about the unique regional diversity and identities of maize and the farmers. Nuestro Maíz is a rural development project coordinated by a regional producer group with ties to a larger national producer organization. The project adds value to local criollo maize, making masa (maize dough) through an industrialized process, which is then made into tortillas that are sold in the communities where the farmers live. The Michoacán Centre for Agribusiness was formed after premium organic markets were diluted. It has been part of the creation of a bioregional label that identifies the ecological and cultural traits of food products and builds regional markets for these products through producer and consumer education. Maize has been reclaimed by the Michoacán Centre for Agribusiness, and native landrace varieties are showcased for their qualities and flavor.

The sociocultural, agroecological, political, and economic aspects of each initiative are distinct, representing different approaches to maize conservation and building local maize networks. Itanoní is reviving and reclaiming Mexican culinary practices through its focus on making tortillas and *antojitos* using maize landraces, typical Mexican ingredients, and traditional cooking processes. The owners of Itanoní articulate the importance of these culinary practices, linking their use and promotion with agrodiversity conservation and food sovereignty. The work of Itanoní is localized to the state of Oaxaca, the four farmers involved, the owners of the restaurant, and the staff and customers. Nuestro Maíz is also committed to using Mexican maize and working with landrace varieties. Their primary focus, however, is not on culinary practice but on rural economic development and creating sustainable livelihoods for campesinos. Nuestro Maíz is working nationally with producer groups that represent larger numbers of farmers. Nuestro Maíz has developed a project model that has been adapted across Mexico. The Michoacán Centre for Agribusiness is an example of a regional labeling scheme that enables farmers to emphasize the qualities of their products to eaters who are not directly connected to production processes. The center is creating local and regional diversified markets for the farmers they work to support. Labeling initiatives have proliferated across Europe and North America and have become important means of conveying information through the marketplace. The Michoacán Centre for Agribusiness is promoting maize landraces in order to diversify agricultural systems and improve subsistence crops.

Itanoní, Nuestro Maíz, and the Michoacán Centre for Agribusiness politicize, engage with, and reveal different aspects of agricultural and dietary transformation in Mexico. The practice of making tortillas, once a daily household activity in much of Mexico, has been largely replaced by tortillas made from maize flour (as opposed to masa, or maize dough) (Lind and Barham 2004). The culinary history of making tortillas at home and the history of the tortilla industry reflect the story of class, race, gender, development, and modernization in Mexico.

Historically, women were the *tortilleras*, tasked with making tortillas for their families. As rural and urban women entered the workforce, it was increasingly difficult for them to maintain the domestic practice of making tortillas, which could take up to three hours each day. New technologies became available to alleviate the hard work of grinding masa with a metate, a stone grinding board used with a large pestle that crushes the maize into flour.

A grinder that ground soaked maize revolutionized domestic life in the 1970s. This technology enabled people to continue the *nixtamal* process—a Mesoamerican culinary invention that transforms the nutrient content of maize to enable nutrients to be better absorbed by the human digestive system—and have fresh, homemade tortillas. The masa grinder is still a common community tool in villages across rural Mexico. Meanwhile, in the urban centers, women who could afford to pay domestic help to make tortillas by hand clung to the laborious practice of using the metate, claiming that the taste was superior (Pilcher 2005).

Commercial tortilla production, since the mid-1970s, has slowly replaced masa with ground maize flour, which is nutritionally inferior (Rivera et al. 2004). The informal market has responded to the Mexican people's ongoing appreciation for maize products made with masa. *Tortilleras* (women selling tortillas and other maize snacks) can be found in all markets and sometimes outside of supermarkets, offering "homemade" tortillas. As people's consumption practices become integrated into the global economy and they shop more frequently at supermarket chains (Chavez 2002), the purchase of tortillas made from maize flour has increased.

Nuestro Maíz, Itanoní, and the Michoacán Centre for Agribusiness, in different ways, are catering to, and attempting to nurture, urban eaters that appreciate Mexican maize, a trend that has connected culinary and nationalist discourses in Mexico since the 1940s (Pilcher 1998). Jeffrey Pilcher describes how urban interest in regional Mexican culinary specialties is related to the emergence of nationalism and Mexican identity. The initiatives are recovering the *nixtamal* process in their tortilla production. In the case of Itanoní, this is being done through small-scale technology and traditional methods. In the case of Nuestro Maíz, this is being done through reclaiming the industrial *nixtamal* process. The campesinos that are working with the Michoacán Centre for Agribusiness highlight the *nixtamal* process as one example of how their products reflect local culinary and agricultural traditions.

In the era of free trade, corporate globalization, and GM corn, it is interesting to find evidence of the reemergence of traditional practices in the tortilla sector. Itanoní, Nuestro Maíz, and the Michoacán Centre for Agribusiness are not alone in their promotion of local maize. Another manifestation of the "tortilla discourse," it is not difficult to find neighborhood *tortillerías* advertising that they use "100 percent maíz mexicano." Even larger corporations use images of the milpa—a concept that also has strong cultural symbolism in Mexico—in their advertising.

Elizabeth Fitting (2006, 2011) describes the political uses of culture in maize movements and networks, a cultural politics of maize that is found at the Itanoní Tortillería, Nuestro Maíz, and the Michoacán Centre for Agrobusiness. She proposes, "Rather than relying on a notion of maize culture that is insulated from wider economic, political and social processes, we might ask in what ways maize agriculture is a cultural practice and how such practices and meanings are transformed by trade liberalization, GM imports, and cuts to rural funding" (25). *Corn Meets Maize*, by taking an intimate look at three local maize networks, sheds light on these practices and politics. Challenging fixed notions of tradition, the initiatives both elucidate and problematize the notion of an authentic Mexico grounded in its maize culture (Fitting 2006, 28). Fitting continues,

> I have suggested that while any culture of corn is a historical product—as the practices and values of peasants are historically constituted through dynamic engagement with wider processes—its conceptualization is also an ideological project of the late capitalist, neo-liberal present (Roseberry 1989, 223). The official government and pro-GM position also employs a reified notion of corn culture, but in support of a neo-liberal agenda. (2006, 29)

Building on William Coleman's idea of transworld agriculture, this book suggests that the initiatives are both embedded within and contesting neoliberalism. The three initiatives are fraught with asymmetries of power, but Verschoor compellingly suggests that the "novel associations between previously unconnected entities attain, once connected, a force of their own. . . . Instead of extinguishing moral and political fires science, technology and economics generate uncertainty and even outright ignorance. But ignorance, uncertainty, or a lack of consensus does not necessarily lead to paralysis. On the contrary, as in the case of opposing GM corn, uncertainty opens up a range of new moral, ethical and political spaces" (2007, 52–53).

Notes

1. The following discussion on food citizenship is adapted from an article published in the *Geographical Review* in 2004, "Tending Cultural Landscapes and Food Citizenship in Toronto's Community Gardens."

2. This section is adapted from an article I published in *Local Environments* in 2008, "Local Food Networks and Maize Agrodiversity Conservation: Two Case Studies from Mexico."

3. Fowler and Mooney distinguish *variety* and *landrace* (a term used later in the following way. "A cultivated plant variety is a distinct, named, rather uniform, modern creation also referred to as a cultivar. . . . Landraces are usually more variable, less

distinct and less uniform. . . . A landrace may express tremendous variation in a single field, days to maturity, even pest resistance" (Fowler and Mooney 1996, xv).

4. Agrodiversity conservation strategies have been developed and implemented by the following international institutions: the Convention on Biological Diversity, the Food and Agriculture Organization, Consultative Group on International Agricultural Research, the International Maize and Wheat Improvement Center, and Slow Food, to name a few. Seed saving and exchange programs, community seed banks, and diversity gardens represent examples of local, regional, and national efforts to conserve agrodiversity.

5. *Nature* later retracted support for the article after criticism from pro-biotechnology scientists. This sparked controversy and debate amongst the scientific community.

6. The literature that speaks most directly to a biocultural perspective emerges from the anthropology and ethnobiology disciplines. Darrell Posey, with his landmark report for the United Nations in 1999, "Cultural and Spiritual Values of Biodiversity," made the links between cultural diversity and biodiversity explicit, paving the way for the UN Declaration on the Rights of Indigenous Peoples passed in 2008. Luisa Maffi, in a recent issue of *Resurgence* magazine, explores the notion of the biocultural as it relates to indigenous peoples' rights, but agricultural biodiversity is marginalized within the broader discussion of biodiversity.

7. The network In Defense of Maíz, according to Silvia Ribeiro, is a coalition of local and transnational environmental and social justice organizations working to educate Mexicans on the impacts of GM corn in Mexico and to lobby the government to prohibit GM corn to be grown in the country (pers. comm., January 15, 2005).

8. I am simplifying the science for argument's sake.

9. Non-shattering refers to a method of seed dispersal common in non-domesticated plants. To disperse, the seed head explodes, sending seeds farther than if they dropped close to the plant. Non-shattering variants of wild plants are central in the process of food plant domestification and enable geneticists to trace evolutionary sequences that resulted in modern cereal crop varieties (Fowler and Mooney 1996).

2

Milpas, Markets, and Movements

Maize in Mexico

> For the independent petty commodity producer, the seed was the alpha and the omega of agricultural life. As seed, it is the beginning of the crop production process, and as grain it is its endpoint. But because the seed is a living and reproducible package of DNA, the endpoint of one cycle of production merely sets the stage for the next. Seed is grain is seed is grain; the option to produce or to consume is there in each seed. And even in consumption there is the element of reproduction. In growing food crops, the farmer may provide for the farm family's means of subsistence. Growing a forage or silage crop provides the means for reproducing livestock (and therefore motive power, food, and fertilizer). Growing a fallow or nitrogen-fixing crop provides the means for reproducing the fertility of the soil. And, of course, seed in the form of grain is a commodity that can be sold for the cash to purchase items of all sorts. Upon the seed depends ultimately the capacity to reproduce a large part of the farm operation. And in control over the seed is a measure of real independence. The seed is the biological nexus of farm-level production.
>
> —Kloppenburg 2004, 37

Introduction

Maize in Mexico is a highly politicized food item with deep cultural and spiritual meaning, intricately connected to Mexico's precolonial and colonial history, as well as to culinary and agronomic resistance (Lind and Barham 2004). Tortillas made from maize are a staple part of the Mexican diet. The

PHOTO 2.1
Maize seed in cold storage at the National Institute for Forestry, Agricultural and Livestock Research's Oaxaca center.

shifting politics of maize in Mexico has been called the "tortilla discourse" and encompasses nutrition, agriculture, trade, and urban food security policy as it has changed and been adapted over time. As Jeffrey Pilcher states in *Que Vivan los Tamales: Food and the Making of Mexican Identity*,

> The tortilla discourse correctly recognized maize as the root of self-supporting communal life, and thus a barrier to modernization, although for cultural rather than nutritional reasons. Nevertheless, the ultimate incorporation of peasants into the national economy came not through the elimination of maize, but rather through its commodification. Once maize had changed from subsistence crop to market commodity, *campesinos* had no choice but to follow it into the modern world. (1998, 97)

As Pilcher describes, Mexican farmers have followed maize into the modern world. An examination of these paths illustrates the ways that policy processes, social relations, farmer management practices, culinary traditions, cultural significance, and civil society organizations (among others) are entangled and interrelated in the story of maize in Mexico. Just as Mexican farmers have followed maize into the modern world, maize has historically shaped agricultural and culinary practices both in Mexico and around the world, galvanizing commodity markets and social movements.

The history of maize production and consumption in Mexico has been explored from innumerable perspectives. Geographers, anthropologists, agronomists, and economists have told the stories of this history, weaving together the agricultural practices, economic policies, and cultural traditions related to maize to illuminate theoretical and empirical points. In this chapter I contextualize the story of maize and corn in Mexico using three diagrams as my guide.

These diagrams describe the various networks within which maize and corn are enmeshed. This is an effort to articulate the numerous actors, the layers of association and assemblage, and the "patchwork quilt" (Peña 1998) of ecologies, markets, and social movements that maize touches. The networks articulated below are incomplete: they are meant to be illustrative not definitive. The purpose of this chapter is to reveal the political ecology of maize and illustrate how the milpa intersects with broader political and economic processes.

The diagrams provide an entry point to begin the discussion about nature-culture, global-local, and market-movement enmeshment. The representations are visual descriptions of maize networks. Following the work of Mark Lombardi (Hobbs 2003), an artist who depicts financial stories and corporate scandals, the drawings are meant to represent the "narrative structures" of maize and corn networks.[1]

I start with a representation of the milpa and the factors affecting agrodiversity. I then build on this diagram to illustrate the interconnections between the milpa and transnational policy-making processes. In the second diagram I describe the socioecological and the political economic interactions between the milpa and transnational policy processes such as NAFTA (North American Free Trade Agreement) and the Convention on Biological Diversity. The third diagram illustrates the social movement actors working "in defense of maize." These diagrams articulate the multiple actors in the maize network including policies, technologies, other crops, the geological region, and farmer management practices. They represent a poststructural approach to understanding the dynamics of maize in Mexico by examining three different, partial narratives, each shedding light on different aspects of the story.

William Coleman coins the term *transworld agriculture* to describe

> the degree to which globalizing processes produce new global transnational spaces by observing shifts in *extensity*, the degree to which cultural, political and economic activities are "stretching" across new frontiers to encompass the "world"; *intensity*, changes in the magnitude and regularity of interconnectedness; and *velocity*, changes in the speed of global interactions and processes. Shifts along these dimensions should lead to increasing enmeshment, that is, the level of interdependence of the global and the local. (Coleman 2003, 3)

The diagrams illustrate the extensity, intensity, and velocity of increasing enmeshment between global-local maize ecologies, economies, cultures, and policies.

The actors in these networks have been galvanized through the "neoliberal corn regime" (Fitting 2006) and the "maize debates" in Mexico. Through these diagrams and the accompanying contextualizing stories, I focus on the ways that maize permeates global-local economies and cultures as well as the cultural politics of the social movements that have come together to defend maize production in Mexico in the face of neoliberal policies. Through examples of translocal maize movements and networks, I demonstrate how maize has been used as a cultural symbol and how maize continues to re-create Mexican cultures and landscapes, as it, in turn, is re-created by the everyday practices of farmers, cooks, markets, and social movements.

By focusing simultaneously on everyday practice and local-global politics, this chapter speaks to the global nature of maize, the interrelationships between maize and people, and the creation of new moral, ethical, and political spaces in the neoliberal landscape. The social movement proposals outlined at the end of the chapter suggest an alternative possibility for maize, one where global policies are challenged by the everyday practices of cultivation and consumption. These everyday practices, in turn, are linked to transnational organizing efforts. The challenge of moving forward with these proposals is formidable, and it is unclear what the future holds for small and medium maize farmers in Mexico. Maize agrodiversity is in their hands.

Maize: Agricultural Artifact

The Milpa

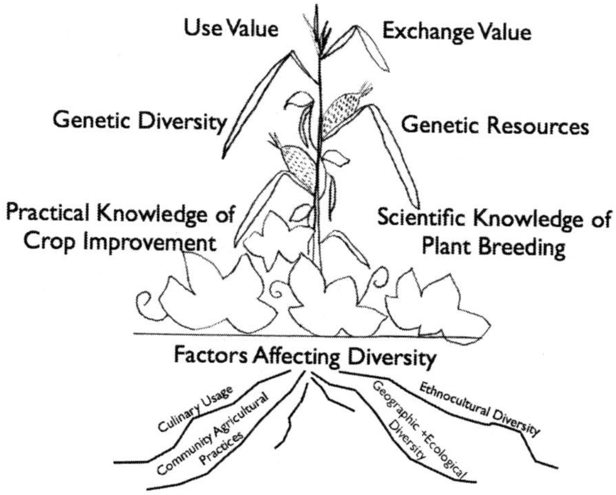

FIGURE 2.1
Diagram A, the milpa

Figure 2.1, diagram A, depicts the milpa—an agroecological system developed in Mesoamerica to cultivate maize and other crops in an agriculturally and nutritionally complementary way. The diagram illustrates the symbiotic relationship between the maize, beans, and squash plants. The intermingling roots show the connection between culinary usage, community agricultural and cultural practices, geographic and ecological diversity, and ethnocultural diversity.

The milpa is a polycrop, and usually includes maize, beans, squash, and *quelites* (weedy greens). Often avocado, tomato, melon, chili, sweet potato, amaranth, and other crops are also grown in a milpa. The roots of the milpa diagram represent the factors affecting food crop diversity. The milpa is shown as embedded in the natural-cultural dynamics of diversity. These dynamics reflect conflicting notions of diversity—the practical knowledge of farmers that has resulted in maize diversity in contrast to scientific conceptions of agrodiversity in plant breeding laboratories and seed banks. These conflicting notions of agrodiversity are explored in this section.

Milpa cultivation is based on agroecological principles. The milpa does not depend on agrochemicals—intercropping enables biological control of insects and diseases; legumes fix nitrogen for the heavy nitrogen demands of maize; squash plants act as a mulch to suppress weeds and conserve soil moisture. The three principal crops—maize, beans, and squash—cycle nutrients efficiently within the system (Altieri 2000; Gliessman 2006). Native landraces and plant varieties are generally used in the milpa, conserving crop genetic resources (Brush and Perales 2007). Crop and landscape diversity is high, promoting biodiversity of insects, soil micro-organisms, birds, and other animals. In and around the milpa subsistence food crops, forage crops, construction materials, flowering plants, and animals can be harvested or hunted. Local agroecological knowledge is promoted and conserved through the milpa, and farmers actively preserve genetic diversity through their production practices (Altieri 2000; Gliessman 2006; Brush and Perales 2007).

The milpa is nutritionally complementary as well. The nutrients that maize lacks—for example, several amino acids—are provided by the beans. The nutritional deficiencies of beans, in turn, are provided for through the maize. The other crops in the milpa provide vitamins and fats (Bressani 1990). Nutritionally, the milpa is an incredibly successful invention and illustrates the potential synergies and dynamism of sustainable agroecological systems.

In Mexico, like maize, the milpa is a cultural symbol that draws meaning from the deeply embedded importance of rural livelihoods and the *vida campesina* (rural life). Guillermo Bonfil Batalla, author of *Mexico Profundo: Reclaiming a Civilization*, describes the interrelationship between the milpa, work, diet, community fiestas, and cosmology.

> Our categories, of Western origin, do not exist in these (indigenous) cultures. In Indian cultures, the conception of the world, of nature and humankind, makes quite different kinds of actions seem equally necessary. For example, a propitiatory ceremony for a good growing season may be as important as the proper selection of seeds for planting. There is a unity of human beings and the natural world, which is the reference point for human knowledge and abilities as well as for work, the specific way of obtaining sustenance. This unity is also present in human plans, in the capacity for imagining as well as observing nature, in the willingness to have dialogue with it, in human fears and hopes faced with forces beyond human control. (1996, 27)

Factors Affecting Maize Diversity

Rafael Paczka identifies four factors contributing to the tremendous diversity of maize in Mexico: culinary usage, agricultural practices, geographic and ecological diversity, and ethnocultural diversity (Paczka 2003, 127). Following, an explanation of each of these factors speaks to the biocultural agrifood relations of milpa cultivation and its relationship to a number of social and cultural practices.

Culinary Usage

Simultaneous to the physical evolution of maize in Mesoamerica was experimentation with numerous culinary uses for maize. The tortilla is just one example of the way maize is transformed into food and drink. Traditional uses of maize link agrodiversity to culinary practices. Culinary use also links agrodiversity to gender, and women's role in agrodiversity conservation. As Kimberlee Chambers and Janet Momsen (2007) have illustrated in their study of gender and maize diversity in the Bajío of Mexico, men and women have different knowledge related to maize. Men identified the maize varieties growing in the fields. Women, on the other hand, identified varieties grown in kitchen gardens surrounding the house. In Chambers and Momsen's study, women's and men's knowledge differs in regards to the identification of varieties grown, seed selection, and opinions about the division of labor. This gendered knowledge is not only related to culinary usage of maize, but includes all aspects of maize production practices. As men migrate out of rural communities to work in the United States, women are responsible not only for kitchen gardens but also for the fields (Chambers and Momsen 2007). The gendered domain of agroecological knowledge shifts in relation to the social and economic context, in this case the out-migration of men to the United States for employment.

Community Agricultural and Cultural Practices

Maize diversity is managed in the field by farmers who carefully isolate and select varieties. Neighboring communities trade and share seeds, resulting in new hybrids and varieties (Perales et al. 2003). Populations of maize, despite this mixing, are kept distinct through rigorous seed selection practices in the field. This combination of mixing and isolation (or selection) plays an important role in maize agrodiversity. Cultural events such as holidays, weddings, and births involve different maize specialties. In *Zapotec Science: Farming and Food in the Northern Sierra of Oaxaca*, Roberto González describes the meaning of maize.

> In the Rincón maize has a soul. Indeed, the customary Rincón Zapotec view of nature is remarkably different from the conventional view in industrialized societies, to the extent that the earth, the rivers, and individual crops such as maize are seen as animate objects and are personified. . . . In the Rincón

> humans are engaged in a reciprocal relationship with these entities, often using saints and mountain spirits as intermediaries or advocates. . . . Thus maize farming in the Rincón involves activities conducted by people enmeshed in complex reciprocal relationships with their families, kin, neighbours, fellow villagers, saints, spirits, and the earth itself. The very embeddedness of maize in these intricate networks has made it an essential part of *campesino* lives that seems unlikely to be given up soon. (2001, 129)

In rural Catholic churches across Mexico, maize is left as an offering to statues of saints around planting and harvest season. This is an example of how maize agrodiversity is intimately connected to community agricultural, culinary, and spiritual practices illustrating the interrelationship and inseparability of nature and culture in the milpa.

Geographic and Ecological Diversity

Maize diversity corresponds to geographical variation or ecological heterogeneity (Toledo 2001). It is not only the macroecological diversity of Mexico's varied topography that contributes to maize agrodiversity. The diversity across a farmer's land also contributes. For example, a farmer might have several plots of land where different varieties of maize are grown depending on agroecological conditions. Farmers choose specific maize varieties to grow in instances of drought or excessive rains. The productive area surrounding the farmer's house represents another ecological niche and is managed and planted differently than plots farther away. The existence of niche environments across regions and scales has led to the conservation and improvement of maize diversity over time. In the case of maize agrodiversity, scale and niche—micro and macro—are particularly relevant.

Ethnocultural Diversity

Ethnocultural diversity in Mexico contributes to maize agrodiversity, as do the contributions of individual communities, families, farmers, and women. By paying attention to the culinary qualities of maize in the field and through seed selection, specific traits and characteristics are favored. Different communities and families grow specific varieties and colors of maize for tortillas (white), animal forage (yellow), snacks (purple), and so on. These differences tend to be variations in the same genetic population, leading to differentiation and variety within maize populations. In many communities certain individual farmers have special interest in agrodiversity and are seen as seed keepers and guardians of maize diversity.

González (2001) describes, in detail, the farming and culinary practices of Talea, a Zapotec community in Oaxaca. Many of the practices are common throughout Mesoamerica, with small, but important differences dictated by culture and ecology. For example, the Talean campesinos describe soils and periods of climatic change in great detail, reflecting their intimate knowledge of ecology and weather patterns.

Maize diversity continuously evolves under the management of small-scale farmers and is shaped by strategies to preserve genetic diversity. In this sense, maize agrodiversity is not fixed or static, but can best be described as a flow of genetic diversity (Paczka 2003, 125–127) that is directly related to culinary usage, community-based agricultural practices, geographic and ecological diversity, and ethnocultural diversity.

The Dynamics of Diversity

Reflecting the management practices of the farmer and community, as well as the local ecology of the farm field, maize agrodiversity expresses astounding genetic variability. For approximately nine thousand years, farmers have been selecting maize seed in their field and storage houses to express traits that include disease resistance, color and taste, grain quality, tolerance to acid or saline soils, resistance to drought, freezes or strong winds, and variable harvest time.

Maize is an open-pollinated crop. Maize pollen can be spread easily by the wind and has been found kilometers from the field in which it is grown. For this reason, maize is said to be "promiscuous." Maize's tendency is to revert from varietal diversity as selected and managed by farmers to varietal homogeneity through cross-pollination. Farmer management practices—such as the act of carefully sorting through and selecting kernels with certain traits from year to year—has ensured ongoing diversity and distinction between landraces. The open-pollinated, promiscuous nature of maize ensures gene flow among landraces that strengthens the genetic diversity. The counteracting farmer management practices manage the gene flow to ensure specific qualities and characteristics of particular landraces are maintained (Mann 2005, 219).

Bellon and Berthaud provide more detail about the farmer management practices that characterize traditional maize agricultural systems and preserve maize agrodiversity (2006, 5). These systems are typically landscapes with multiple maize populations. Four particular practices manage these populations and maize agrodiversity. First, seed saving affects the genetic structure of the population through the selection of specific qualities and characteristics. Second, seed exchange between farmers means that new populations are incorporated, new characteristics are introduced into the system, and the dynamics of diversity (gene flow) are managed. Third, farmers introduce landraces from other regions every few years to improve their maize populations by enhancing heterozygosity (genetic variation). Fourth, farmers incorporate both foreign landraces and improved varieties into their farming systems. As they select the seed of these varieties from year to year, the introduced varieties become "creolized," or mixed, combining characteristics of all varieties and landraces. These practices speak to the sophisticated practical knowledge of farmers.

Jack Kloppenburg contrasts this practical knowledge of crop improvement with the science of plant breeding in *First the Seed: The Political Economy of Plant Biotechnology* (2004). In order to understand the dynamics of diversity, I draw on Kloppenburg's analysis of how social relations have shaped seed production as a means of capital accumulation. Using the example of corn breeding in the United States through the nineteenth and twentieth centuries, Kloppenburg illustrates shifts in seed production that culminate in the science and politics of plant biotechnology in the twenty-first century. Kloppenburg's analysis of this history provides a framework with which to understand the dynamics of maize diversity in Mexico. The chart in table 2.1 articulates the key differences between practical knowledge and scientific knowledge.

TABLE 2.1

Practical Knowledge and Crop Improvement	*Scientific Knowledge and Plant Breeding*
Seed as whole organism	Seed as genetic components
Focus on seed selection	Focus on seed hybridization
Crop improvement as art	Plant breeding as science
Genetic diversity as a public good	Plant germplasm as a private resource
Seed as a commons that is used	Seed as a commodity that is exchanged

The chart contrasts the notions of practical knowledge related to crop improvement and scientific knowledge related to plant breeding. A key difference is the conception of the seed as a whole organism versus the seed as a compilation of genetic components that can be isolated and selected through breeding practices. Kloppenburg further articulates this as the difference between crop improvement as an art and plant breeding as a science. Crop improvement, explains Kloppenburg, focuses on seed selection by farmers who closely observe the traits and characteristics of the plants expressed in the field and, in the case of maize, on the cob. This careful selection results in the management and improvement of genetic diversity by farmers in their fields. The practical knowledge of seed selection and crop improvement is based on a conception of genetic diversity as a public good, and seeds as a commons that are *used* by and shared between farmers, illustrating use-value as opposed to exchange-value.

Plant breeding, on the other hand, focuses primarily on generating hybrids to improve yield, although other objectives such as disease resistance are also important. Plant breeding and the process of creating hybrids result in seeds that need to be purchased by farmers, resulting in a loss of control of the agricultural process, dependency on purchased seed, and alienation from the seed selection and crop improvement process. The seed, in this case, is a commodity considered for its *exchange-value* and derived from plant germplasm patented as a private resource.

Several key points emerge from Kloppenburg's work that I would like to carry forward into the following chapters. The first is the distinction between genetic diversity and genetic resources. Genetic diversity can be seen as a flow—between farmers, communities, fields, and plants. Genetic resources, through international agreements and transnational corporations, are conceived of as germplasm that can be controlled and manipulated. The second is the distinction between economies and cultures of use versus economies and cultures of exchange. Genetic diversity, in the initiatives I explore in the following chapter, is used—by farmers, agronomists, public crop improvement specialists, cooks, and eaters. The use-value of genetic diversity challenges the dominant conception of genetic resources as a commodity with exchange-value for the life sciences industry. These distinctions will be explored in the following sections.

The Local/Global Importance of Maize Agrodiversity

As previously illustrated, maize formed (and continues to form) the spiritual, cultural, and economic center of Mesoamerican society. Since the colonization of the Americas, maize has been an important crop globally, as well. Maize is adaptable to diverse environmental and agronomic conditions, which has led to its relatively easy spread and acceptance around the world. Associated with the characteristic of adaptability is a parallel characteristic that has led to the broad commercial uses and applications of corn. Corn has proven extremely adaptable to industrial technologies and has been a key input for many industrial manufacturing processes beyond the food sector, including energy (biofuels) and plastics. Corn hybrids have provided incredible productivity gains for farmers, and in comparison to other cereal crops, the corn seed industry is the most commercial seed industry globally, generating tremendous rates of return for investment in hybrid corn research (Kloppenburg 2004, 91; Wise 2007, 4). The improvement, hybridization, and genetic modification of corn seed has proven extremely lucrative for commercial plant breeders and seed companies, as growers have to purchase new seeds each year to reap the high yields delivered by hybrid seed.

Kloppenburg notes three features of the commercial plant breeding methods that ensure profits for plant breeders and seed companies. First, the process is too complex for farmers to replicate in their fields. Second, seed cannot be saved without drastic yield reductions. Third, the inbred lines that are used to create the hybrids means seed companies have proprietary control over the hybrids (2004, 99).

In order to understand the history of corn breeding and the ascent of commercially controlled hybrid corn seed, I briefly examine some of the terminology used. The language is essential to understanding the debate surrounding the conservation of maize agrodiversity, and Kloppenburg's (2004) fascinating history of hybrid corn offers some clarity on this terminology.

The complex interactions between plants and people over time resulted in genetically variable populations within one species called *landraces* (referred to as *criollos* in Mexico). These genetic variations offer different responses to diseases and environmental conditions. *Germplasm* is the hereditary material contained in the cells of plants that express this variation. This diversity has become known as *plant genetic resources*, the raw ingredients for plant breeding and biotechnology.

In 1901 the rediscovery of Austrian geneticist Gregor Mendel's work laid the foundation for commercial *hybridization* with the understanding of how genetic traits express themselves. The art (and unpredictability) of crop improvement suddenly became predictable. "Instead of selecting from the diversity in nature," states Kloppenburg about plant breeders, "he saw germplasm as something to be molded in a predictable fashion" (2004, 69). Before this time, and through the early part of the twentieth century, hybridization implied cross-breeding two varieties of a plant. From 1930 to 1940 plant breeders began seeking germplasm with specific traits, signaling a shift from crop improvement to plant breeding and a reductionist shift from the whole organism to the genetic components in the form of germplasm. Kloppenburg describes,

> There is an even more compelling reason to examine closely the historical choice of breeding methods in corn, for the use of hybridization galvanized radical changes in the political economy of plant breeding and seed production. There is a crucial difference between open-pollinated and hybrid corn varieties: Seed from a crop of the latter, when saved and replanted, exhibits a considerable reduction in yield. Hybridization thus uncouples seed as "seed" from seed as "grain" and thereby facilitates the transformation of seed from a use-value to an exchange-value. (2004, 93)

Kloppenburg makes several key points about the history of plant breeding science. It relies on an asymmetrical flow of genetic diversity from the global south to the north, which is then sold as genetic resources back to the south in the form of hybrid varieties. It is a *social* process shaped by public and private investment in plant breeding. Public funding and research undergirds scientific research and development. Different plant breeding technologies have different commercial applications, and the technologies with the highest potential for profit have dominated, at the expense of other technologies.

To illustrate the distinction between genetic diversity and genetic resources, as well as how the tension between use-value and exchange-value plays out in Mexico today, I offer an example from the Nuestro Maíz crop improvement program.[2] Yolanda Salinas Moreno, a geneticist working for INIFAP (the National Institute for Forestry, Agricultural and Livestock Research) at the University of Chapingo[3] describes how their crop improvement partnership with Nuestro Maíz is working to improve criollo varieties by crossing them with open-pollinated hybrid varieties.

> One of the main problems with criollo maize is that the plants are very tall. When it is windy, the plants are blown over. In the high valleys of Mexico, this is the principal problem with the criollo varieties. We are improving the criollos by incorporating gene flow from hybrids that aren't so tall. This conserves a large part of the criollo germplasm. You aren't going to plant hybrids in these areas. The environment is too harsh and the farmers don't have the resources [irrigation and fertilizers] to grow hybrids. The goal is to conserve their own varieties. Leave them as criollos, but stop them from falling over. (pers. comm., February 24, 2006)

Gene flow between criollos and hybrids is common, but the criollo characteristics and traits are selected in the field, maintaining while improving genetic diversity. This complicates static definitions of agricultural biodiversity and suggests that the metaphor of gene *flow* is more appropriate than a static conception of diversity.

With the reductionist shift from crop improvement to plant breeding, the germplasm and plant genetic resources of Mexican maize were systematically collected through various publicly funded programs, precursors to the Green Revolution. In 1943 the Rockefeller Foundation founded the Mexican Agricultural Program and the United States began amassing maize germplasm. In 1970, after a maize blight decimated U.S. farmers' harvest, the Green Revolution was institutionalized through the founding of the Consultative Group on International Agricultural Research. The Green Revolution intensified the flow of plant genetic resources from the diversity-rich south to the diversity-poor north, where commercial breeders relied on publicly funded germplasm accessions, collected in seed banks, to develop hybrids that were then sold to farmers in the north and south, along with technological packages of herbicides, pesticides, and fertilizers (Kloppenburg 2004).

In the 1970s, following several decades of systematic collection and documentation of maize in Mexico and after the institution of the Green Revolution, there was a renewed interest by both Mexican agronomists and the international genetic resources institutions in the relationship between maize agrodiversity and campesino livelihoods (Paczka, pers. comm., January 25, 2006). This interest had three roots. First, a blight attacked maize in the United States, resulting in a search for maize varieties that might be resistant to the blight. Second, maize improvement programs in Mexico were seeking to increase the efficiency of small-scale farmers in rural Mexico, recognizing that the genetic diversity these farmers were using was not adequate and was not represented in the germplasm banks.[4] Third, there was a resurgence of interest in maize agrodiversity and ethnobotany due to the work of professor Efraím Hernández Xolocotzi[5] at the University of Chapingo (Paczka 2003, 125). The work of Hernández Xolocotzi inspired a new generation of research on maize agrodiversity in Mexico, and a renewed interest in the importance of campesino agricultural practices among researchers.

This renewed interest in maize agrodiversity led to the development of in situ (in the field, managed by farmers in partnership with researchers) and ex situ (seed bank) conservation strategies. If maize agrodiversity is most appropriately conceived of as a flow of farmer knowledge and genetic diversity, the efficacy of seed banks as a way to conserve maize diversity must be questioned. Seed banks have been criticized by indigenous communities as a technical and scientific solution that leads to dependency and alienation (Ribeiro 2004, 9), serving plant breeding scientists instead of supporting farmers to manage diversity in their fields. Also, the collection of maize samples has been shown to be ineffective over time. Salinas Moreno provides the following example:

> The *hala* race is a very large maize, with large cobs. It was adapted to a harsh environment in Nayarit. It isn't found there anymore. We have it catalogued in a seed bank, but these collections were made 50 years ago, and in 50 years the environment has evolved. So, when you take the seed from the bank to where it was collected and plant it, it isn't adapted to the environment. This is really serious. (pers. comm., February 24, 2006)

Another missing link is the ecological knowledge of the campesinos who grew this variety, chose the seed from season to season, and planted the seed in relationship to the other plants in the milpa. The separation of the seed from the local ecology and cultural practices related to milpa cultivation limits the utility of seed banking.

The Red de Maíz del Sistema Nacional de Recursos Fitogenéticos (the Maize Network of the National Phytogenetic Resources System) was formed in 2003 to assess agrodiversity conservation efforts and advise Mexican policy makers. According to Paczka there is a divide in the network among those who support ex situ and in situ conservation strategies. The issues are complex. For example, many in situ programs have not been contiguous over time due to funding constraints. These shifting priorities reflect the states' position on agricultural biodiversity and mirror neoliberal policies in the way state support of these programs ebbs and flows. In addition, farmer management practices and culinary uses are increasingly recognized as important to food plant diversity (see Chambers and Momsen 2007). Attention to maize agrodiversity conservation shifts over time, with particular moments catalyzing a deeper interest in these issues, engaging a wider number of actors. Paczka summarizes, tongue in cheek, these issues as genetic discourses versus genetic resources (pers. comm., January 25, 2005). Here, Paczka refers to the shifts in agrodiversity politics and interests over time, reinforcing my distinction between genetic diversity and genetic resources.

Paczka describes the motivations behind this renewed interest in maize agrodiversity as a recognition of the relationship between maize, people, and the broader ecosystem, in particular the importance of understanding and protecting rural traditions that form such an important part of Mexican identity. Mexican agronomists and maize researchers are increasingly aware of campesinos' global contribution to knowledge related to maize production, and their important role in conserving maize diversity for food security globally (Paczka 2003, 125).

These shifts in agrodiversity politics can also be interpreted as confrontations between knowledge systems. Gustavo Esteva, in his article "Hosting the Otherness of the Other: The Case of the Green Revolution," provides examples of how knowledge systems overlap and intersect through his description of the work and approach of three agronomists working on maize agrodiversity (or campesino production) in Mexico (Esteva 1996). Through the stories of the careers and approaches of these agronomists, Esteva illustrates how knowledge systems are perpetuated, respected, disregarded, and/or marginalized. Moving beyond a recognition of how these knowledge systems "operate in dialectical interaction" (272), Esteva draws on the concept of radical pluralism in an attempt to avoid "the reduction—to each other—of the knowledge systems" (249). Using Hernández Xolocotzi's approach to maize conservation as an example of radical pluralism, Esteva describes,

> In clear contradiction with everything he learned in school and with most of the beliefs of his colleagues, he suspects that the world's essence is of a plural nature. He is perhaps suspecting that the radical diversity or pluriformity of

> plants, which he found a long time ago, may be in the society of men in the form of radical pluralism. Through corn, his main obsession, he obtained a clear insight into a very wide and mysterious spectrum of radically different attitudes toward the prodigious plant. I think that he finds it both silly and dangerous to try and reduce all of them to only one vision, to one classification, to only one parameter. (272)

From the choices each farmer makes in the field, the community relations that shape production, in situ and ex situ genetic resource conservation, and the Convention on Biological Diversity to U.S. agroindustry and agricultural policy, the milpa represents the place where nature and culture, policy and practice intersect. Devon Peña describes spaces such as the milpa as

> an epistemology of place [that] open[s] new avenues for the expression of the social and cultural practices of local, or situated, knowledge. By "situated knowledge" we understand the dialectical relationship of humans to the biogeographical properties of place in which, over generations, the local culture accumulates a vast reservoir of knowledge dealing with the ecological limits of life in a specific locale. Norgaard (1994) describes this relationship as a "co-evolution" of culture and environment in which local communities become part of a "patchwork quilt" of diverse cultural and ecological landscapes. (1998, 11)

It is increasingly recognized that small- and medium-scale farmers are the guardians or keepers of agrodiversity, and that their livelihoods are threatened. The practical and intimate knowledge of these farmers is intricately connected to spiritual and cultural practices. Neoliberal globalization threatens these livelihoods and practices. Vandana Shiva summarizes this threat:

> The fragmented linearity of the dominant knowledge disrupts the integration between systems. Local knowledge slips through the cracks of fragmentation. It is eclipsed along with the world to which it relates. Dominant scientific knowledge thus breeds a monoculture of the mind by making space for local alternatives to disappear, very much like monocultures of introduced plant varieties leading to the displacement and destruction of local diversity. Dominant knowledge also destroys the *conditions* for alternatives to exist, very much like the introduction of monocultures destroying the very conditions for diverse species to exist. (1993, 12)

In the maize story, this epistemology of place stretches beyond local ecologies and cultures that are intimately enmeshed, and is just as directly informed by transworld agricultural policies, civil society movements, and corporate food systems. To further explore the dynamics that shape maize agrodiversity, the following section describes the interrelationships between the milpa and transworld agriculture.

The Milpa and Transworld Agriculture

> Local farmers who have developed maize genetic diversity for several millennia in Mexico's varied agroecosystems continue to provide a critical environmental service. Every year they perform the task of selecting seeds and blending genetic diversity that is vital for agriculture not only in Mexico but on a global scale. Because of their capacity to preserve maize genetic diversity, this crop is one of the most important in the world. This also explains why maize can grow at sea level or 2000 meters, in acidic or alkaline soils, under conditions of drought, with long and short periods of maturity, at different heights and wind resistance, and with the capacity to withstand several local pests. But the negotiators who crafted NAFTA did not consider these variables in the equation and concluded, using the single datum of yield comparison, that local farmers were inefficient and should give way to their more able colleagues from North America. (Nadal 2006, 34)

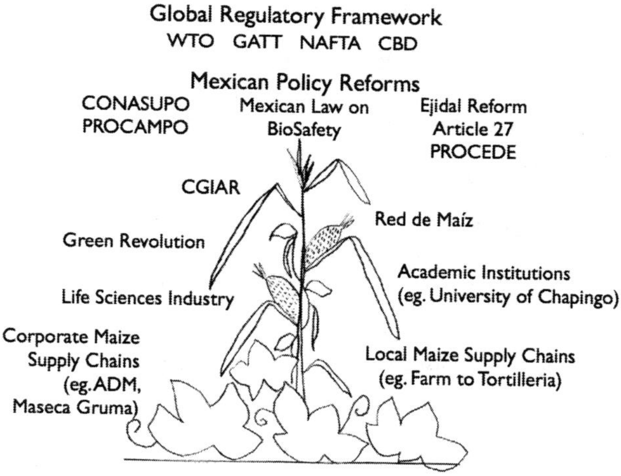

FIGURE 2.2
Diagram B, the milpa and transworld agriculture

Elizabeth Fitting (2006) describes the overarching contemporary policy processes in which the milpa and maize production in Mexico are enmeshed as the "neoliberal corn regime."[6] This section, beginning with figure 2.2 (diagram B), describes the history and impact of neoliberal policy in relation to the milpa (as described in the previous section), maize production, and consumption in Mexico. The "neoliberal corn regime" has resulted in an unprecedented change in Mexican rural life and the agricultural sector, threatening the livelihoods of small and medium farmers and maize agrodiversity (see also Pechlaner and Otero 2008). The milpa is embedded in a network of national and international policy processes. These connect maize to a transnational space, stretching cultural, political, and economic activities beyond the milpa, increasing enmeshment and interdependence.

Neoliberal Reform of Agriculture and Food Policy

Mexico's embrace of neoliberal policy can be traced to 1982 and the Miguel de la Madrid government. When Mexico defaulted on its foreign debt payments in 1982, a neoliberal policy framework and structural adjustment program were imposed by international lending institutions, characterized by privatization, deregulation, and trade liberalization (Fitting 2006, 16). Under the administration of Salinas de Gortari (1988–1994), the agricultural sector underwent reform. These reforms radically restructured the Mexican countryside by eliminating food and agricultural-input subsidies, liberalizing imports of agricultural goods, reducing the scale of publicly funded farm credit and technological assistance, and privatizing communally held *ejido* lands (Preibisch, Herrejón, and Wiggins 2002, 69). In this section I trace the implementation of neoliberal policies and agricultural reforms in Mexico. In the following section I discuss how these policies and reforms have impacted small- and medium-scale farmers and agrodiversity.

The amendment of Article 27 in the Mexican constitution in 1992, known as the *ejidal* or agrarian reform, had profound impacts for small and medium farmers and agrodiversity (Yunez-Naude 2002). The *ejido* reform gave flexibility to the land tenure structure and redefined property rights in rural Mexico. Communally owned *ejido* lands were dismantled, and renting and selling this land was legalized. Furthermore, the reform legalized the ownership of this land by corporations, foreign and national. The purpose of the privatization of *ejido* land was to enable entrepreneurial peasants to invest in their property and increase the economies of scale of their production to facilitate integration into the market economy. In reality, many campesinos in Mexico lacked the capital to pursue these investment opportunities. The law has promoted private investment in the countryside. Through this mechanism, U.S. and Mexican agribusinesses have been able to obtain land for export-oriented production, displacing domestic supply chains.

Ejido reform undermined maize diversity (Dyer and Belausteguigoitia 2006). Victor Toledo describes how community-based tenure such as the *ejido* "provided a protective shell for the functioning and evolution of resource management systems responsive to local ecological and biodiversity conditions" (2001, 474). Smaller landholdings and the increased numbers of households farming on the *ejido* favored the use of maize landraces over high yielding hybrid varieties. The dismantling of *ejidos*, the increase in private, corporate investors in the rural sector, and the consolidation of land holdings has favored commercial corn production using hybrid varieties over subsistence maize production using landraces.

Hector Díaz Castellano, an organic farmer who supplies criollo maize to Itanoní describes how the dismantling of the Mexican *ejido* system has changed farming practices in his community and the agricultural landscape.

> Before PROCEDE[7] the land was communal. Everyone who lived in the *ejido* worked together, for the good of the *ejido*. No one could abandon a parcel of land because the law didn't allow it—the community would cultivate it. Now, if someone abandons their land for two or three years it just sits there and no one in the community can use it. . . . When the land was communal the harvest was abundant because the work was well organized. It was much better than it is now, cultivating our individual parcels. (pers. comm., February 15, 2006)

Ejido reform is but one way that Mexico's agriculture and food policies impact maize agrodiversity. Agriculture and rural development policy has waxed and waned historically in its support for Mexico's campesino farmers and the *vida campesina*. In postrevolution Mexico, the PRI (Institutional Revolutionary Party) administration pursued economic policies based on self-sufficiency and national food security. The period between 1935 and 1982 was characterized by government intervention in the agricultural sector. Mexico's small and medium farmers were provided with crop price supports for basic grains, agricultural inputs were subsidized, and the government participated in the grain processing sector. The rural vote was linked to policies that subsidized subsistence food production such as the Companía Nacional de Subsistencias Populares (CONASUPO) program created in 1961 to guarantee the price for basic grains and to distribute subsidized foods (Fitting 2006; Preibisch, Herrejón, and Wiggins 2002; Appendini 1996). CONASUPO controlled the national market for maize as well as maize supply to rural areas through supply stores known as *diconsas*. Programs like these represented the recognition by politicians of the importance of the rural vote and campesino livelihoods (Antal 2007a, 15).

In her analysis of neoliberal agricultural reform in Mexico, Kirsten Appendini describes two overlapping, complementary and contradictory policy frameworks that shape maize agrodiversity and, more broadly, the Mexican countryside. The broad projects of "modernizing, agricultural support programs, and 'globalizing'" trade described by Appendini help to explain the policy shifts implemented by the Mexican state as it attempts to integrate the Mexican agricultural sector into global markets (Appendini 1996).

The modernization project involves policies and programs that affect the majority of rural farmers: 3.2 million farmers, who feed 12.5 million rural people and grow maize on plots of land five hectares or smaller, are affected by the policies of the modernization program (Vera Herrera 2004 in Antal 2007a). These farmers lack the economic strength to participate in global markets. The goal of the modernization project is to stimulate productivity in the Mexican countryside through a policy platform supporting small-scale peasant farmers.

A parallel policy approach, the globalization project, implements policies that encourage production for export markets and expand Mexico's fresh fruit and vegetable exports through intensive, agroindustrial production practices. Approximately five hundred larger-scale farmers are engaged in the agricultural production that is favored by the globalization program. Overall, the modernization and globalization projects, through asymmetrical policies, unfavorably support a minority of larger-scale, export-oriented farmers in Mexico, while further marginalizing small- and medium-scale farmers.

These policy programs have resulted in the restructuring of core agricultural policies. The CONASUPO program described previously was dismantled in the early 1990s and replaced in 1993 by PROCAMPO (Programa de Apoyo Directo al Campo, Direct Rural Support Program). This resulted in the elimination of credit subsidies and fixed prices for basic grains, replacing them with a crop neutral subsidy (a direct income transfer allocated per hectare) to compensate farmers for decreasing prices resulting from trade liberalization. With the dismantling of CONASUPO the state withdrew participation in the basic grain market. This withdrawal is significant for several reasons. First, it impacts national food security as prices for basic grains became subject to the volatility of the global market, impacting both farmers and eaters. Second, the direct income transfers available through PROCAMPO are not sufficient to replace the subsidy in the form of fixed prices. There are limited funds that are difficult for the most marginalized farmers to access. The removal of a fixed price for basic grains, combined with the flood of corn imports from the United States, means that the price of maize is below the Mexican cost of production. This results in small- and medium-scale farmers no longer being able to sell their maize on the national market for a profit.

In January 1994 Mexico signed NAFTA, consolidating the neoliberal agenda. NAFTA established tariff rate quotas as a transitional mechanism to eliminate barriers to trade between the United States, Canada, and Mexico. These quotas were applied to products considered sensitive under the trade agreement, including corn. A fifteen-year phase-out program was established, which ended in January 2008, to complete the liberalization process. Quota tariffs were agreed upon by the three participating states, and the quotas grew each year until free market access was reached (Yunez-Naude 2002).

Maize and dry beans were granted special status as sensitive items under NAFTA. This safeguard clause protects products such as basic grains in Mexico and can be used as a "countervailing mechanism." The mechanism can be put into place when an increase of imports represents a "considerable menace" to the sensitive product. The mechanism also allows Mexico to suspend tariff reductions or use the 1994 base tariff. Despite these clauses under NAFTA, Mexico only charged minimal tariffs (1–3 percent) on any crops from 1994 to 2000 (Yunez-Naude 2002). In 2004 the Mexican Congress charged higher tariffs: 72.6 percent for white corn in 2004, 54.5 percent in 2005, 36.3 percent in 2006, and 19.2 percent in 2007 (Korves 2007). The early uncollected quotas resulted in foregone revenues that could have been applied to the sustainable development of the Mexican rural sector (Wise 2007). This refusal to use the countervailing mechanism provided through NAFTA meant that imported maize from the United States flooded the Mexican market.

The flood of highly subsidized corn from the United States to Mexico after NAFTA was implemented in 1994 meant that maize prices for Mexican farmers dropped 48 percent between 1990 and 1998 (Preibisch, Herrejón, and Wiggins 2002, 69). Esteva explains the impact of asymmetrical farm support policies and imported corn on local farmers.

> Local farmers have always had a local market for maize. There has always been a demand in rural areas for locally grown maize. This stimulates production because it is an income source. Small farmers can't enter the national market, but they can receive an income from the local market. As an example, local maize sold for four pesos a kilo. Then the DICONSA, an official agency, sells maize from Michigan for two pesos a kilo. This clearly diminishes local farmers. Even though people don't like imported maize (because it is a variety used for livestock, not for human consumption) the difference of half the price is huge. People can't ignore this. Because of their economic circumstances they are forced to buy the imported maize and use it for tortillas. (pers. comm., February 17, 2006)

Esteva uses this example to stress the aggressive neoliberal policies the government has pursued to undermine campesino subsistence production since 1982 (also see Fitting 2006, 16). Subsistence farmers in Mexico grow maize for their families in the milpa but are not outside of the market. Many sell small amounts of their surplus grain locally, providing small- and medium-scale farmers with some income. This connection between the milpa and the local market is threatened by imported grain available for subsidized prices.

The 3.1 million farmers who grow maize in Mexico (Cevallos 2006; Toledo 2001) have been deemed inefficient in comparison with highly subsidized corn farmers in the United States. Indeed, average yields per hectare are very low—1.8 tons in Mexico compared with 9 tons per hectare in the

U.S. corn belt (Nadal 2006). However, small-scale farmers growing for their personal consumption produce just under half of the national demand for maize in Mexico. Another quarter of the demand for maize in Mexico is supplied by medium-scale farmers who sell in the national market. The rest is imported (Antal Fodroczy 2004).

Pre-NAFTA government involvement in the maize sector guaranteed prices for farmers that were above global market prices. For example, when NAFTA was negotiated in the early 1990s, U.S. corn cost $110 per ton, while Mexican farmers received $240 per ton (Antal 2007a, 3). More importantly, Mexican agricultural policy protected small- and medium-scale farmers from commodity price fluctuations in the global grain market. Furthermore, guaranteed prices for farmers were coupled with food subsidies for eaters.

The price volatility of corn has impacted farmers, eaters, and the tortilla sector. After NAFTA was signed, while prices for basic grains plummeted, the price of tortillas soared. Between 1994 and 1999 tortilla prices rose 483 percent (Nadal 2000). The Tortibonos program was a tortilla voucher program geared toward urban eaters implemented in the 1980s to protect urban eaters from price increases due to neoliberal agricultural reform (Appendini 2001). This program was eliminated in 1999. As the *Economist* reported in 1996, "Normally, big business mourns the passing of any handout in Mexico; but a free market in tortillas could mean a bonanza for several big firms" (*Economist* 1996, 73). The decline in maize prices following NAFTA did not translate into reduced prices for Mexican eaters. According to Nadal (2000), oligopoly in the corn importing and processing sector allowed the dominant agribusiness actors such as Archer-Daniels-Midland and Maseca/Gruma to reap the profits from the lower corn prices (Wise 2004, 9).

Indeed, the elimination of urban food subsidies such as the Tortibonos program has meant increased centralization of the tortilla industry, with a drop in the nutritional quality of Mexico's most basic food item (Pilcher 2005, 241). Tortillas produced with corn flour instead of with nixtamalized maize cannot be digested as easily, resulting in fewer of the nutrients available for humans. Tortilla subsidies for the urban poor have been intricately connected to corporate control over, and the industrialization of, the tortilla sector. In the 1950s and 1960s the Mexican State Food Agency provided inexpensive corn to the tortilla industry, which then sold tortillas to the public for fixed prices (Pilcher 2005, 240).

Grupo Gruma (a Mexican multinational corporation), under the tortilla and corn flour brand Maseca, has dominated the maize market since the 1970s. Although nutritionally inferior, Maseca justified the shift by centralizing and standardizing production and enriching the tortillas with proteins and vitamins (Pilcher 2005, 241). Through joint ventures with Archer-Daniels-Midland (a U.S. multinational corporation), Gruma controls the majority of the corn-tortilla chain in Mexico. These corporations are vertically integrated along the supply chain and control almost half the maize in Mexico, purchase the majority of the national maize supply, import corn from the United States, and own grain storehouses, process corn into flour for tortillas, and market well-known tortilla and corn products (Antal 2007a, 15).

Manuel Chavez (2002) documents increasing concentration in the Mexican retail sector, which has also impacted local maize markets. Before the mid-1980s none of the main domestic retailers found in Mexico today were established. Following Mexico's signing of the GATT (General Agreement on Tariffs and Trade) in 1986, strategic alliances began to form in the retail/commercial sector. Post-NAFTA, the annual growth of supermarket sales was 7.3 percent, and by 2005 supermarkets controlled 60 percent of the food market (Chavez 2002, 506). This rapid concentration of the retail sector since NAFTA has shifted purchasing and consumption practices, severely impacting smaller retailers, fresh food markets, and independent supermarkets.

Despite the increasing concentration in the retail food sector, small-scale, neighborhood-based tortilla production thrives in Mexico. Tortilla sales at supermarkets and convenience stores only represents 2.5 percent of total national sales (S. González 2008). However, the use of corn flour, replacing the *nixtamal* process, has increased. In 1978, 78 percent of the maize used in the tortilla sector in Mexico was made into *masa* through the *nixtamal* process. By 1989 it was 64 percent (Appendini 2001, 195). By 2005 the split between masa (*nixtamal*) and flour (*harina*) was half and half (Velázquez 2005).

Yet the capitalist penetration of the corporate corn supply chain is incomplete. Corporate tortilla brands compete with neighborhood *tortilleras* and "Marias" selling tortillas, *tlacoyos*, *tlayudas*, and other "typical" maize products, sometimes right outside large supermarket chains. Vizcarra Bordi (2006), in her ethnography of Marias (Mazahua and Otomia women who sell tortillas on the streets of Mexico City), describes how tortillas and other traditional maize products have "undergone hybridization processes . . . as a reaction to the confinement of homogenization." She continues, "The inclination or preference to eat tortillas made by hand with nixtamalized (lime-cooked) maize, has not only been a response to a taste strategy by eaters but, also, to economic strategies (of indigenous and subsistence farmers seeking access to the market), in which multiple power relations are concealed" (97). This culinary preference supports a local market of small maize farmers, masa suppliers, neighborhood *tortillerías*, as well as the Marias described by Vizcarra Bordi.

Pilcher, in his article "Industrial Tortillas and Folkloric Pepsi: The Nutritional Consequences of Hybrid Cuisines in Mexico," states, "The modernization of food production has been instrumental in drawing campesinos into the market economy, but it has done so in a halting and incomplete fashion, creating culinary versions of what Nestor García Canclini termed 'hybrid cultures'" (Pilcher 2005, 235). Changes in tortilla production at the household and industrial level can be linked to changes in food policy, corn policy, and appeasing rural voters, but also speak to the endurance of culinary and taste traditions despite these policies.

Support for the majority of Mexico's small-scale farmers continued to be eroded over the Vicente Fox administration (2000–2006) as policies and programs to support export-oriented agriculture were favored and direct subsidies through PROCAMPO were cut. Funding was increasingly channeled into rural enterprise development efforts that did not reach the majority of small- and medium-scale farmers. The Felipe Calderón administration (2006–2012) continued to reform agricultural policy, refused to contemplate demands to renegotiate NAFTA, and followed through with Mexico's NAFTA commitment to dismantle all corn tariffs in January 2008. In late 2006 the price of tortilla flour almost doubled over several months, leading to public outcry and "tortilla" protests. In January 2007 Calderón capped the prices of corn products. January and February 2008 saw massive protests and rallies calling for the renegotiation of NAFTA. On February 1, 2008, a huge rally was held in Mexico City with estimates of two hundred thousand participants. Most recently, the tortilla discourse manifested itself through tortilla protests in response to the rising price of corn on the world markets due to the agrofuel boom. Because corn forms the foundation of their diet, Mexican eaters were hit particularly hard by these global price spikes.

A compounding factor has been the introduction of genetically modified corn to Mexico. This is a result of planting imported corn from the United States. GM (genetically modified) corn crossed with native maize landraces resulted in concern about Mexico's maize agrodiversity (Cummings 2002). Public concern about GM corn in Mexico was sparked in 2001 with the publication of a paper in *Nature* (Quist and Chapela 2001) describing the contamination of Mexican maize landraces by imported GM corn. This contamination, although challenged by the biotechnology industry, was confirmed by independent tests done by indigenous communities in Oaxaca (see Fitting 2006, 2007; Antal 2007a, 2007b; Massieu Trigo 2007).

The agrofuel boom has been the most recent chapter in the maize story. The U.S. announcement in 2006 to increase corn-based ethanol production as a strategy to respond to rising oil prices led to corn futures spiraling out of control. This resulted in a sharp increase in the price of corn and sent ripples throughout the food system, causing food prices to rise around the world and leading to widespread food riots.

The Supporting Global Regulatory Framework

The national-level reforms and policies detailed previously are consistent with the global agricultural regulatory framework outlined in the GATT and by the World Trade Organization (WTO). The GATT and WTO institutionalize these trends affecting food system development, as outlined by Geoff Tansey and Tasmin Rajotte in their book *The Future Control of Food: A Guide to International Negotiations and Rules on Intellectual Property, Biodiversity and Food Security* (2008). The global regulatory framework consolidates power in the food system. The ability of corporate players to influence prices, reduce competition, and set standards is growing. Local and national markets are subsumed by regional and global markets, and the number of actors involved in the food system is diversifying.

In 1987 Mexico signed the GATT, but agricultural reforms did not take place until the early 1990s, when agricultural trade liberalization was extended through the Uruguay Round Accord. In 1995 the WTO was formed, institutionalizing binding dispute settlement mechanisms backed by sanctions. The WTO brought agriculture into the trade regime for the first time, introducing rules on sanitary and phytosanitary standards and intellectual property through the Agreement on Trade Related Aspects of Intellectual Property. The reforms previously described are consistent with the transnational policies of the GATT and WTO to decouple subsidies from crops and set import taxes to protect sensitive crops.

One other global agreement must be mentioned in this discussion. In 1992 Mexico ratified the Convention on Biological Diversity. The National Biodiversity Strategy developed by Mexico as part of this ratification does not deal thoroughly with agrodiversity in Mexico. Directly related to maize agrodiversity, however, is the Mexican Biosafety Law. In 1999 an Inter-Ministerial Commission on Biosafety was created by the Mexican government to oversee the regulation of GMOs (genetically modified organisms). Between 1999 and 2005 the gaps in regulatory policy were exposed by a number of civil society organizations aiming to address the impacts of NAFTA and transgenic corn introgression (Fitting 2006, 17). After a contentious policy-making process, Mexico passed a law on biosafety and GMOs in December 2004 (Fitting 2006, 18). The law has been controversial, as it does not ban the use, sale, import, export, and farming of GM corn, instead outlining the process through which to obtain permits for such activity and rules requiring the labeling of imported GM corn. In 2009 the first GM trials were approved.

As part of the Biosafety Law, a special government program to protect maize agrodiversity and other species for which Mexico is the center of origin was also established (Lotter 2004). Critics have pointed out that the regulatory process does not go far enough. In particular, critics are concerned about an allowance that states shipments of corn containing less than 5 percent of GMOs do not need to be identified. Shippers who unknowingly transport GM corn into the country are not held accountable (Fitting 2006, 18). Mexican regulation of transgenic food and crops is intricately enmeshed in environmental, agricultural, sanitation, health, and security legislation at the national level, and reflects Mexico's commitment to the WTO, NAFTA, and the UN Convention on Biological Diversity.

Although it is outside the scope of this chapter to fully discuss the global regulatory framework supporting the neoliberal corn regime in Mexico, it is useful to outline some key points. These conflicting and contradictory policy processes (between rural development, national biosafety legislation, and global agreements) make it extremely difficult to enact regulatory frameworks that comply with the numerous commitments made at the national and international level. The challenge is one endemic in policy structures. State bodies are fragmented into silos, addressing different aspects of agricultural policy.

Despite the weaknesses and limitations of the Mexican law on biosafety it is an important achievement. Timothy Wise suggests that there is further policy space available for protecting maize agrodiversity in Mexico, even in the face of the complete liberalization of corn since January 2008. The space requires Mexican politicians to make use of policy instruments available through existing international agreements. Space exists within NAFTA to renegotiate parts of the treaty (as opposed to rewriting the whole agreement) and address the environmental concerns related to biosafety and biodiversity. Specifically, opportunities are created through GATT to protect agrobiodiversity as an environmental good and service. The Cartagena Protocol also provides mechanisms (related to labeling) to regulate GM corn, but these mechanisms have not been fully implemented through Mexico's Biosafety Law. Wise states, "Any government committed to developing policies more supportive of traditional maize farmers would find a great deal of room to move" (Wise 2007, 17), illustrating how Mexico is not taking advantage of the policy space available within current agreements. Wise describes what is possible within the current policy context, but the mechanisms that could be drawn upon are weak (Friedmann, pers. com., April 24, 2008).

Three examples are given by Wise and briefly noted here. First, through NAFTA Mexico could make the case to impose countervailing duties on U.S. corn imports and increase its own agricultural subsidies to small-scale farmers, mirroring the highly subsidized sector in the United States.

Second, under the GATT, Mexico could justify protection of maize agrodiversity as an environmental good. Third, Mexico could establish more rigorous labeling of GM corn imported from the United States through the Cartagena Protocol (Wise 2007, 9–19). Even, however, if Mexico did shift its policies to support campesino farming systems and maize agrodiversity, the agreements limit the extent of this support and could put Mexico in a position to be challenged by trading partners.

The multiple international agreements to which Mexico is a signatory represent a tangle of obligations related to trade, agriculture, and biodiversity. Although NAFTA, GATT, the WTO, and the Convention on Biological Diversity offer limited avenues for protecting maize agrodiversity and rural livelihoods, what is lacking is the political will to find the policy space necessary to defend maize. Many groups, including Via Campesina, the international peasants' organization, call for the dismantling of this global regulatory framework and replacing it with "a decentralized model of production, processing, distribution and consumption [that is] controlled by the people and communities themselves and not by transnational corporations" (Via Campesina 2008). As is noted in the conclusion of Antal, Baker, and Verschoor (2007), civil society and indigenous organizations are using the counter-hegemonic position they have gained through their campaigns over the past few years to demand that the government take advantage of the policy space available (82). In the current political climate it seems unlikely that this will work, given the entrenchment and commitment of the Peña Nieto government to neoliberal policies.

The previous sections have sketched the history of neoliberal agriculture and food policy reform in Mexico and reviewed the global regulatory framework that supports these policies. The result of this global-local entanglement has been a dramatic shift from use-value to exchange-value economies and cultures of maize and corn. The milpa is not place-bound, but shaped by these policy processes. The asymmetrical policies I described earlier, favoring corporate interests at the expense of rural livelihoods and local urban food security, have resulted in economic and social inequities and externalized environmental costs of global corn trade. These issues are briefly reviewed in the following.

Impacts of Neoliberal Policy on Campesinos and Agrodiversity

The Mexican government argument in support of free trade has been that opening borders and focusing on export-oriented growth will create jobs, in particular for the rural and urban poor, lifting many Mexicans out of poverty. The Mexican government's own statistics illustrate that Mexicans have not been lifted out of poverty in the thirteen years since NAFTA was signed. In fact, poverty increased and deepened between 1994 and 2008. Over 60 percent of people living in extreme poverty in Mexico live in rural areas (Council on Hemispheric Affairs 2008). Employment in agriculture and rural wages have shrunk since 1992. Neoliberal policies and the intensification of poverty in rural Mexico have had a major impact on the Mexican agricultural sector (see Fitting 2007; Preibisch, Herrejón, and Wiggins 2002). For the purposes of this chapter, I will focus on two broad impacts that affect small- and medium-scale farmers and agrobiodiversity. The first is migration and the second is changes in maize production.

Migration to the United States has increased since NAFTA came into effect. This is a result of losing access to land, debt related to high input costs, and other impacts of neoliberal policies. Rural areas account for 25 percent of the Mexican population, but 44 percent of migration (Council on Hemispheric Affairs 2008). In her research in the Tehuacán Valley, Fitting describes how international policies and neoliberal reform have led to out-migration to the United States (Fitting 2007, 2011), impacting agricultural practice. Households in this region increasingly rely upon remittances to subsidize their farming livelihood—to purchase agricultural inputs and basic necessities that are beyond a meager farming income. The younger generation of migrants (in their twenties or younger) were not able to describe to Fitting how to cultivate maize in detail. Fitting illustrates how "the fate of Mexican maize biodiversity not only depends upon clear and enforceable biosafety regulations but also upon policies that support and enhance sustainable agriculture and small-scale Mexican maize cultivators" (2006, 23). Fitting's research documents the transnational livelihood strategies of Mexican farmers, as well as the ways policy and practice intersect to transform landscapes and food economies and cultures over time.

Hector Díaz Castellano, a small-scale farmer from El Trapiche, Oaxaca, is frank about the importance of remittances he receives from his children who have been working in the United States since 2003, as well as the cultural changes that result from transnational livelihood strategies. Only one of four children has remained in El Trapiche and works on the farm. One son has financed the building of another house on the property. When asked if he could imagine a Mexico without maize production Díaz Castellano answered,

> Yes, I think so. The relationship we have with our neighbor [the United States] has undermined the small farmers. It is much easier to buy maize than to grow it. In our communities, we are losing the tradition of making tortillas. I am almost sixty years old and in my life I have seen a big change. Before we made tortillas by hand on a wood-fired comal. Now most of our utensils are plastic. This doesn't give our food any flavor. . . . I'm not saying what we have now is bad. For example, a housewife in a community doesn't have to make tortillas or burn wood if she has money to buy tortillas. The problem is that to get this

money the majority depend on people who are away [working in the United States]. (Díaz Castellano, pers. comm., February 15, 2006)

Kirsten Appendini also describes how migration combined with neoliberal policies is changing the rural landscape. Supported by remittances sent from relatives abroad, rural communities are increasingly seen as a refuge from migration, places where parents and grandparents continue to cultivate the milpa while the younger generation build country homes as a place to escape from their demanding jobs in the United States (pers. comm., January 24, 2006). Both Fitting and Appendini's research, as well as my interviews in Mexico, suggest that maize production in the milpa is under threat from increasing migration of the younger generation, and an aging agricultural population.

Deborah Barndt (1999) describes the NAFTA migration cycle. Not only do Mexicans migrate to the United States and Canada to work as temporary migrant workers, they also migrate within Mexico to work for export-oriented agribusiness. The majority of seasonal migrant workers within Mexico are indigenous families from southern states where *ejidal* reform and debt have left them without land for subsistence agriculture. These seasonal workers follow the harvest, leaving their communities and milpas. Barndt documents the gendered and racialized nature of these jobs. For example, mestizo women are employed as packers, while indigenous people are employed as pickers. As women enter the workforce, the practice of making tortillas changes due to increased pressures on women's time.

Despite what some have called the "assault" on Mexican campesinos, subsistence production and maize production have remained stable since 1994 (Preibisch, Herrejón, and Wiggins 2002). This paradox speaks to the many extra-economic reasons that people produce maize in the Mexican countryside. Maize is valued for cultural and spiritual reasons—for example, its taste, use, and agroecological contribution in the field. As rural people become more economically insecure, they grow maize for its contribution to household food security (Escobar Moreno 2007). Victor Toledo describes the productive strategy of Mexican campesinos in the following way:

> Since peasant production is based more on ecological exchanges than on economic exchanges, indigenous households tend to adopt survival mechanisms that guarantee an uninterrupted flow of goods, materials, and energy from nature. . . . Thus, they carry out a nonspecialized production based on the principles of diversity of landscapes, resources, species, genes, and as a consequence, of productive practices. This multiuse strategy, through which indigenous peasants maintain and reproduce their productive systems, constitutes an ecologically valuable characteristic that tends to conserve natural resources

and that buffers peasant households against both market fluctuations and environmental changes and hazards. (2001, 476)

Gisele Henriques and Raj Patel, researchers at Food First/Institute for Food and Development Policy, further describe the paradox (Henriques and Patel 2004, 5). Small-scale farmers in Mexico lack options and resources, making it very difficult to consider changing their production practices. Instead of diversifying their crops, farmers tend to intensify their production practices, putting additional pressures on the land and labor. Not only is the land under cultivation used more intensively, but farmers apply more pesticides and fertilizers to increase yields. More land is sought to cultivate, placing pressure on the surrounding environment.

Farmers recognize maize as a risk-adverse crop—one that is safe to grow with predictable harvests—another reason maize production has remained stable. In addition to being a reliable crop, maize is also a staple food in the diets of rural Mexicans. If small-scale farmers have limited money to purchase food in the market, they will continue to grow their own maize. Because different varieties of maize are used in community and family celebrations, rituals, and ceremonies, as well as for distinct dishes, rural families choose to grow specific varieties of maize, many of which are not available on the market. Finally, farmers choose to continue growing maize because there are few viable alternative crops that will fetch a good price at the market. This is because there is wide availability of imported food available for a relatively inexpensive cost.

In addition to the social, cultural, and economic ramifications of liberalization, there have also been environmental ramifications. For example, forests have been cleared to produce more maize, and land under cultivation is being farmed more intensively. Pest infestations such as white flies have caused the devastation of extensive regions under horticultural production (Antal 2007a; Barndt 2002), due in part to the refusal of agribusinesses to let the ground lay fallow periodically.

To summarize, the impacts of the "neoliberal corn regime" in Mexico have been felt all along the production-consumption chain, but most intensely by small-scale campesino farmers. Returning to the notion of transworld agriculture described by Coleman, these shifts reflect an increasing extensity, intensity, velocity, and enmeshment between globalizing processes and local landscapes. The local is infused by the global, illustrating how the milpa and agrodiversity are intricately connected to global processes. Since the implementation of NAFTA in 1994 there has been an unprecedented change in Mexican rural life and the agricultural sector, threatening maize agrodiversity. This has led to a proliferation of civil society efforts articulating the importance of maize in Mexico.

Maize Movements: Civil Society in Defense of Maize

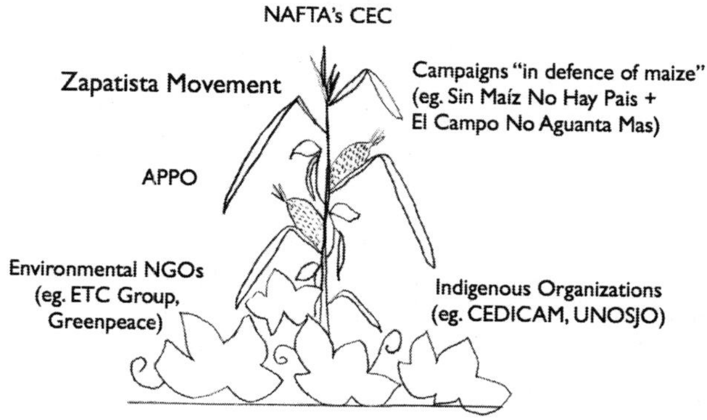

FIGURE 2.3
Diagram C, maize movements

By focusing simultaneously on everyday practice and local-global politics, the groups referred to in figure 2.3 (diagram C), speak to the global nature of maize, the interrelationships between maize ecologies, cultures, and economies, as well as the possibility for creating new moral, ethical, and political spaces in the neoliberal landscape. In a sense, the proposals put forward by these groups and described in this section suggest another possibility for maize, one where global policies are challenged by the everyday practices of cultivation and livelihood and linked to transnational organizing efforts. Small- and medium-scale farmers, along with these civil society organizations, are redefining productivity and agrodiversity in relation to their practical knowledge and agroecological systems, reflecting new and emerging biocultural agrifood relations. The challenge of moving forward with these proposals is formidable, but the translocal networks developed by these movements and initiatives provide an example of how food ecologies and cultures resist, reflect, shape, and are entangled with the story of neoliberalism.

The ongoing liberalization of the maize trade in Mexico along with the GM corn issue reinvigorated a broad number of social movement actors to address various facets of the issue. Harriet Friedmann suggests, in her discussion of the emergence of a new food regime, that social movements play (and have played historically) a critical role in regime crisis and formation (2005a, 231). She documents new alliances that link food and agriculture issues to a whole suite of other issues that include environment, health, energy, and so on (251). These issues are also taken up by agrifood corporations, as can be seen by the increasing attention to quality supply chains (254–260). Social movements have responded to the standardization and neoregulation[8] of quality by "reassert(ing) the deeper meanings of agroecology and community in ways that do not reject markets but rather seek to "multiply niches" as an alternative strategy to the expansion of one set of products and procedures at the expense of all others (Fonte and Boccia 2004)" (Friedmann 2005a, 261). The example of maize in Mexico offers opportunities to understand the ways social movements in Mexico are contesting and shaping the emerging food regime.

Examining the intersection between neoliberal "neoregulation" and the adoption of agricultural biotechnologies, Gabriela Pechlaner and Gerardo Otero (2008) argue that social movements have the potential to "modify, or even derail, the technology's role in individual nations, and, accordingly, (in) the unfolding food regime as a whole" (2). Using North America as a case study, Pechlaner and Otero document resistance to biotechnology, suggesting that these campaigns such as Sin Maíz No Hay Pais (Without Maize There Is No Country) could, potentially, alter the trajectory of the emerging food regime.

This book contributes to these discussions and suggests that by framing the activities of social movements as biocultural agrifood relations, the debate is moved beyond dichotomous expressions of "alternative" and "dominant" food systems, to a more nuanced analysis that contributes to the discussion about how social movements shape regime formation.

The social movement groups described in the following discussion are linked through their concerns about neoliberal policies, food sovereignty, and biotechnology, and are active in transnational campaigns related to these concerns, such as the World Social Forum, Via Campesina, the ETC Group, and Greenpeace. Several campaigns launched by civil society organizations in the early 2000s illustrate the ways that maize galvanizes people to promote, protect, and conserve Mexico's maize culture.[9] These campaigns provide context for the initiatives to follow in the succeeding chapters, illustrating the biocultural agrifood relations related to maize in Mexico, and how the material and symbolic meanings of maize are shifting under the current social, political, and economic context.

The first campaign, Sin Maíz No Hay Pais, was conceived in 2000 as part of an effort to call national attention to government policies that threaten campesino livelihoods and traditional agricultural systems (Esteva and Marielle 2003). The campaign was mounted as an exhibition at the Museum for Popular Culture in Mexico City and was followed by the publication of a book by the same title. The second campaign was launched in 2002 by a group of civil society organizations from regional, national, and international institutions and organizations. The group submitted a request to the NAFTA Commission for Environmental Cooperation (CEC)[10] to examine the impacts of transgenic corn in Mexico.

The Sin Maíz campaign was developed by the Centre of Intercultural Encounters and Dialogue, the Environmental Studies Group, and the National

Directory of Popular Culture and Indigenous Peoples,[11] involving indigenous communities, activists, and academics to call national attention to these policies. According to Esteva, the exhibition was criticized heavily by the federal government, which insisted that Monsanto be included in the story of maize in Mexico (pers. comm., February 17, 2006). Monsanto was not included, but local governments, seemingly under pressure from the federal government to reject a formal association with the exhibit, prevented its circulation around the country. The book and the exhibit at the Museum for Popular Culture in Mexico City were launched in parallel to the CEC process, with many of the supporters of each campaign involved with the other.

The CEC process was launched in 2002 with a petition submitted to the CEC by twenty-eight indigenous communities (primarily from Oaxaca) and three civil society organizations (the Mexican Centre for Environmental Rights, the Union of Environmental Groups, and Rural Studies and Campesino Support).[12] The petition requested an analysis of the impacts of transgenic corn on maize agrodiversity and local ecosystems, an analysis of the impacts of gene flow from GM corn on the genetic diversity of maize landraces, and an assessment of the impact and risk of current and future contamination (CEC 2004, 32). The petitioners also requested that specific recommendations be made to the Mexican government related to regulating biotechnology.

The CEC advisory committee that directed the analysis and report was composed of a wide range of stakeholders—academics, civil society organizations, and representatives from the biotechnology sector. The selection process for the advisory committee was contentious. Critics stated that too many of the members had links with the biotechnology industry. The original group had no representatives from indigenous farming communities, nor any maize specialists. Pressure from the petitioning group led to changes in the committee's make up (Lotter 2004), and the final committee included representatives from indigenous and farmer organizations.

On the 11th of March, 2004, a draft of the final CEC report was presented at a public forum in Oaxaca, Mexico. The forum was extremely well attended. The civil society groups involved in putting together the CEC petition and organizing the Sin Maíz No Hay Pais campaign encouraged hundreds of farmers to attend. Observers of the forum marveled at the coming together of academics, policy makers, indigenous farmers, and activists. The report's key findings and recommendations address gene flow, biodiversity, health, and sociocultural matters.

Of particular interest to many scholars studying the CEC report and its outcomes (Antal Fodroczy 2004; Ruggles 2006) is the report's acknowledgment of sociocultural issues. The report situates the issue of transgenic introgression historically, recognizing the effects of rural transformation, trade imbalance between the United States and Mexico, and the intersection between rural poverty and biodiversity in Mexico. The report acknowledges the deep cultural, symbolic, and spiritual meanings maize has for Mexicans, and asserts that the risk of transgenic corn is linked to these meanings and values (CEC 2004, 23). Also emphasized is the perceived threat transgenic corn poses to "political autonomy, cultural identity, personal safety and biodiversity" (CEC 2004, 23). The Canadian and U.S. official response to the report was critical. Both countries state that key scientists and stakeholders were omitted from the process. The Canadian response notes that a discussion of "existing domestic regulatory approaches and international obligations of the three governments" was ignored. The U.S. response charges that an economic analysis was not applied to the recommendations, and that many are logistically infeasible (CEC 2004, 22).

In its official response the Mexican government requested that the CEC report discuss the "possible benefits of genetically modified maize for the development of agriculture in Mexico" (CIBIOGEM [Commission on Biosecurity and Genetically Modified Organisms] in CEC 2004). The Mexican official response also requested that value judgments be avoided in the document. The following sections from the report's "Conclusions on Sociocultural Aspects" were given as examples of inappropriate value judgments.

> In the regions of maize landrace cultivation, there is a recent cultural memory and political history among the indigenous peoples of perceived inequity and injustice at the hands of Mexicans of Spanish origin, Americans, and powerful elites. . . . Similarly, those who advocate greater use of genetically engineering and unrestricted trade may have vested interests in aspects of scientific and technical development, trade, political influence, or industrial agriculture in Canada, Mexico and the United States. (CEC 2004, 23)

The CEC report was endorsed by civil society organizations such as the ETC Group and Greenpeace, and has been used as a lobbying tool by these groups in efforts to strengthen the Mexican biosafety protocol as it was being developed.

Antal, a Mexican scholar researching biotechnology policy in North America, makes several interesting points related to the CEC report. One novel aspect of the report was that it put the social and cultural risks associated with transgenic corn on par with the environmental and economic risks. The risk discourse was expanded from the environmental and economic risks described by participating scientists to include the risks related to losing the practical knowledge of campesinos related to agrodiversity. Indigenous groups participating in the CEC process articulated the risk they face from neoliberal policies and transgenic corn—the risk of losing their identity and livelihoods. This risk is linked to notions of food sovereignty—the right to a sustainable livelihood based on farming, the right to participate in local markets, and the right to save seeds. This contrasts with scientific notions of risk limited to food and biosafety (Antal 2007a). Drawing on Kloppenburg (2004), this reflects an emphasis on the use-value economies and cultures of small- and medium-scale campesinos in contrast to the exchange-value global economy. Finally, Antal states,

> [T]he main lesson we should learn from the CEC meeting is not what the exact gene is that is being crossed with maize in Oaxaca and to what degree, which was never even clarified anyway. The main lesson is a reflection about the capacity of governments to regulate agricultural production in a rational, legitimate way given the people's enormous distrust and anxiety. The lack of credibility and trust in private and public institutions, whether they be universities, research centers or government agencies, was clear. It was shown that the peasantry has not participated in decision making about maize in Mexico, the country with the world's most diverse germplasm, characterized by the large number of small farmers in its agricultural system, by a complex agrarian history and by marked social polarization in which small peasants, who make up two-thirds of the agricultural sector, produce half of all maize output. (2004, 117)

When asked what he thought was the cumulative impact of these civil society maize campaigns, Esteva replied,

> I believe that these struggles—Sin Maíz, No Hay Pais, the CEC committee, Itanoní and others—what they are doing is affirming our traditions. The campesino tradition is agrodiversity. Diversity is their tradition and they practice biodiversity in their everyday lives. It is part of their productive scheme. The campaigns affirm these traditions. The effect is beautiful. (Esteva, pers. comm., February 17, 2006)

Linking tradition with everyday practice, Esteva suggests that agrodiversity is part of the economies and cultures of campesinos. What is affirmed is the practice of agrodiversity, not specific varieties and landraces. The skill of crop improvement is part of this practice, as are culinary and community uses of maize. It is the everyday biocultural practice of diversity, and fluid notions of "tradition," that are defended by maize movements and alliances. Diversity is dynamic, linked to farmers, communities, and ecologies, in contrast to the static notions of agrodiversity reflected by strategies such as seed banking.

Two other related campaigns deserve brief mention and contribute to this history of the maize debates in Mexico, as told through the stories of civil society organizing efforts. After the CEC process was formally complete, many of the civil society groups involved created the In Defense of Maize movement to continue articulating the cultural and economic importance of maize in Mexico. Their activities have supported ongoing testing for transgenic corn introgression in rural areas, illustrating that the contamination is more widespread than previously thought (CECCAM [(Centro de Estudios para el Cambio en el Campo Mexicano, Center for Studies about Change in the Mexican Countryside] et al. 2003). These campaigns follow on the heels of a coalition that worked throughout the 1990s named El Campo no Aguanta Mas (The Countryside Cannot Bear It Any More),[13] organizing several mass rallies in the historical center of Mexico City to raise awareness of the impact of NAFTA and neoliberal trade policy (Fitting 2006, 18).

Maize and Biosecurity in Mexico: Debate and Practice (Antal, Baker, and Verschoor 2007) traces the biotechnology-biodiversity debate in Mexico, highlighting the various actors involved in the debate and emerging practices. Antal discusses the positions taken by various actors in the maize debates in Mexico. She concludes that two opposing positions related to biosecurity exist in Mexico, the first based on free markets and economic growth and the second based on food sovereignty and social, cultural, and agricultural values of diversity (Antal 2007a, 30). Antal illustrates how the maize debates are transnational on several fronts. For example, the biotechnology corporations are primarily multinational and the environmental and indigenous groups involved in the issue belong to transnational antiglobalization networks. The actors came together through the CEC, a transnational policy process.

Verschoor examines the ways that social movement actors in the debate frame their arguments. He identifies three justifications of these actors (2007, 40–44). First, justifications based on equity and solidarity connect farmers' unions, producer associations, and indigenous groups in Mexico with struggles for food sovereignty around the world. Central to food sovereignty movements are issues of access to land, sustainable livelihoods, the social and environmental costs of neoliberal development, and biopiracy. The second justification used by social movement actors in the maize debates is based on tradition and locality. Implicit to these claims

> is a reference to ancient, local craftsmanship which gives life to local villages and that is in danger of extinction . . . just beneath the surface of the account is a rich array of distinct cultures, traditions, and crafts that are highly differentiated across regions. These domestic types of evaluatory frameworks furthermore accentuate the harmonious, quasi-romantic experience of living off the land—that is, a close connection between people and their area, a relationship mediated by maize. This treasured local area is a valued piece of collective heritage as well as a location of special meaning for the personal biography of those who live there. (Verschoor 2007, 44)

The third justification is based on ecology and *la vida campesina*, illustrated by the claims made by indigenous groups who articulate the ways that their entire way of life—for example, their cultural traditions, agricultural practices, local economies—is threatened by GM corn. I see these three kinds of justifications as interconnected and relational, drawn upon simultaneously by social movement actors.

Verschoor states that fragmentation of the actors in the movement can be interpreted as problematic and indeed may result in the movement remaining marginal. He suggests, however, that the practices of the groups involve opening new moral, ethical, and political spaces.

> Their originality lies not so much in the denunciation of market or industrial forms of reasoning, but rather in the bringing to attention of complex associations between all sorts of beings: beliefs, institutions, politics, regulations, eaters, environments, markets, sciences, traditions, biotechnology, not to mention transgenes . . . the interrelatedness between all humans and non-human participants in the dispute. (2007, 49)

This comment suggests that by focusing on practical knowledge related to agrodiversity and campesino and indigenous livelihoods, social movement actors are moving away from oppositional politics to focus on everyday practice as political. In the following section I suggest that what is interesting about these maize movements is that they are focused on regenerating and creating biocultural agrifood relations. The proposals articulated by these actors illustrate the tensions, contradictions, and potential of these movements as they navigate the neoliberal context.

Proposals in Defense of Maize

The campaigns and actors involved in the maize debates have articulated many proposals to address the GM corn issue, linking agrodiversity to food sovereignty. The proposals address disparate aspects of the issue but have some common themes. In this section I review some of the proposals made, draw out themes, and discuss the tensions. The proposals help to reveal the new moral, ethical, and political space created by civil society in defense of maize—space that reflects the neoliberal context as well as contests it.

RMALC—Mexican Network against Free Trade

RMALC (Red Mexicana de Acción frente al Libre Comercio, or Mexican Network against Free Trade) is a network of civil society organizations working on issues related to globalization and free trade. The organization was initially formed to fight NAFTA and is part of the Social Hemispheric Alliance that emerged from the anti-NAFTA coalition. RMALC was at the forefront of protests against the final dismantling of agricultural tariffs in Mexico under NAFTA. A campaign they launched in January 2008 led to over two hundred thousand people protesting in the streets of Mexico City. On its website, RMALC describes these events: "This is the moment that civil society, all Mexicans, wherever we are in the world, are raising our voices to defend maize. This is defending our sustenance, our sovereignty and our freedom. Because without maize, there is no country. Put Mexico in your mouth" (RMALC 2008).

RMALC had ten demands.

1. Take maize and beans out of NAFTA. Install a permanent mechanism within Congress with which to administer imports and exports of maize and beans (and their derivatives and subproducts).
2. Prohibit the planting of transgenic corn in Mexico. Protect and improve the genetic diversity of Mexican maize and offer incentives to grow native and organic maize.
3. Approve the Constitutional Right to Food in Congress and the Planning Law for Agrifood Sovereignty and Security and Nutrition in the Senate.
4. Fight against monopolies in the agrifood sector. Avoid monopolies and speculation as well as the deceptive publicity of *chatarra* food. (*Chatarra* food means junk food. In this context it refers to industrially produced tortillas.)
5. Promote Mexican maize and its cultural expressions (culinary traditions and symbolic meanings) and register it as soon as possible in the List of Oral and Intangible Human Heritage at UNESCO (United Nations Educational, Scientific and Cultural Organization).[14]
6. Control the price of a basket of basic, staple foods (this refers to the cost of a basic nutritious diet), guarantee supply, and create a strategic food reserve. Promote the consumption of campesino foods and fair trade.
7. Recover the rights of indigenous communities and protect campesino territories and their strategic natural resources.
8. Support more coffee farmers[15] to gain access to international markets and the best prices.
9. Promote forest and rainforest conservation by sustainable management of natural resources by community organizations.
10. Guarantee the principle of gender equality in rural policies, as well as understanding human rights, citizenship, and labor of agricultural workers and migrant workers.

These demands address neoliberal policy generally and the agricultural sector in particular. The breadth of the demands—from renegotiating NAFTA to changing Mexican agricultural policy to recognizing rural livelihoods as a *right*—indicates the collaborative process used by RMALC and the diversity of constituents in their network. The demands make linkages between policy and livelihood, and between rural and urban citizens, as well as environmental stewardship and human rights.

The proposals demand that the global regulatory framework be reworked to accommodate campesinos and recognize the importance of maize, suggesting that the global arena is an appropriate forum through which to do this. The RMALC proposals make demands of the state, demanding a shift away from neoliberal economic policy to food sovereignty. Subsidies for basic grains, food as a human right, incentives to protect agrodiversity, and protection against corporate monopolies in the agrifood sector are recognized as state responsibilities by RMALC. The inevitability of the global regulatory framework is inferred by RMALC and reveals a tension between neoliberal policy reform and the creation of food sovereignty policy. The tension is resolved by thinking about these as fluid processes responding to specific historical contexts, instead of static proposals. Nevertheless, the proposals put forth by RMALC lead to the question, is the global regulatory framework essential or contrary to agrodiversity preservation as a practice grounded in the practical knowledge of crop improvement and use-value economies and cultures?

Forum in Defense of Maize

Silvia Ribeiro, from the ETC Group in Mexico City, reported back from the Second Forum in Defense of Maize in an article published in *Seedling* magazine in 2004. Initially, the groups present at the meeting demanded more testing for contaminated maize in their fields. Ribeiro describes a shift in their demands after they realized the futility of this.

> The communities recognized that what made them vulnerable to contamination were a series of national and international economic and political factors (free-trade agreements, massive migration, cultural and food erosion, urban and rural poverty, etc.). They came to the conclusion that they could only defend maize by defending the wholeness of peasant and indigenous life along with their rights and resources. (Ribeiro 2004, 9)

At the end of the meeting the participating groups declared a moratorium on transgenic plants and stated that they would not plant unknown seeds or cook with unknown maize. This in turn led the groups to state that they would promote local, anollo seed and strengthen local and community seed exchange systems. Other strategies articulated at the meeting included improving local cooking habits; valuing traditions, myths, and ceremonies; and encouraging community planting, harvesting, cooking, and exchange activities. Stories about contamination would be collected and investigated in order to exchange information on abnormalities related to GM contamination. The threats of GM corn would continue to be discussed and the groups would continue with their anti-GM lobbying activities. Finally, the forum proposed to strengthen and expand links with both urban and rural groups working to promote the consumption of criollo maize from local markets.[16]

This focus on everyday practices and the use of maize illustrates a shift from a singular focus on GM corn and a static notion of agrodiversity to a deep critique of neoliberal corporate globalization. This shift profoundly challenges the scientific knowledge of plant breeding, recognizing the practice of agrodiversity as inseparable from community, cooking, ritual, and agroecology. The strategies proposed represent an assertion of intent to reclaim food sovereignty as something that is practiced on a daily basis (Kneen 2011). This shift is an epistemological shift, challenging the regulatory framework on a much different level than the RMALC proposals. The focus on practical, everyday strategies both contradicts and complements the work of RMALC.

Mexican indigenous and civil society groups, through the preceeding contributions, are among movements across the world that are broadening the global debate on GMOs and intellectual property rights. The debate is shifting to reflect new actors who, in turn, insist that agrodiversity is explicitly linked to the sustainable livelihoods of small-scale farmers. This shift—from seed saving and banking to affirming the evolving biocultural agrifood relations of communities—is an example of a shift in public consciousness and collective action strategies related to the cultural politics of agrodiversity.

The proposals reviewed in this section are fragmented and come from disparate voices in the struggle to address biotechnology and biodiversity in Mexico. However, some common themes can be drawn. First, trade policy and the regulatory framework that has threatened the livelihoods of small- and medium-scale farmers and allowed for GM corn in Mexico are challenged. The proposals call for a moratorium on transgenic plants, changes in international trade policy, and a comprehensive food policy that addresses both production and consumption issues. Second, the proposals address the complexity of agrodiversity and the connections between agrodiversity, indigenous livelihoods, and global policies. Third, the proposals are multiscalar, addressing both the everyday practices and more structural issues that impact agricultural biodiversity. Finally the proposals recognize the rights of local communities to define their livelihoods and suggest alternatives that are grounded in equity and justice for both rural and urban citizens.

The proposals focus on the cultural, ecological and economic importance of maize to campesino livelihoods. These issues, however, are connected to a transnational discourse, and through these transnational linkages, Mexican indigenous, peasant, and civil society groups are connected to other place-based struggles in defense of culture and ecology and against corporate interests. The contradictions between the proposals of RMALC and the Forum in Defense of Maize are illustrative of the complexity of the issue and the multi-scalar linkages between milpas, markets, and movements. The way that certain aspects of the cultural and symbolic meanings of maize are used by transnational movements differs from strategy to strategy. Often these aspects of the maize discourse are marginalized, dismissed, or omitted due to rationalist thought in transnational movements. What is interesting in the maize debates, however, is that these symbolic meanings are articulated as important by a diversity of social movement actors, and have been transformed into a strategy for organizing. This represents a post-global, translocal politics grounded in the everyday practice of farming, cooking, and eating. Maize movements both challenge and delegitimize corporate and policy regimes that undermine maize culture (Wittman, Desmarais, and Wiebe 2010, 27–28). Linking ecology and culture—through multi-scalar networks of people, plants, and policies—are biocultural agrifood relations.

Conclusion

This chapter serves to introduce multiple dimensions of the maize and corn story in Mexico. Social movement efforts and campaigns to protect and promote maize agrodiversity illustrate the new actors galvanized by the threat of neoliberal policies to articulate the importance of maize in Mexico. These campaigns are embedded in ongoing historical processes that have shaped politics, culture, and agriculture in Mexico. The integrated ecological and cultural perspectives combined with global-local political and economic perspectives used in this chapter serve as a preliminary understanding of biocultural agrifood relations, a concept that will be elaborated throughout this book.

In the first section of this chapter I illustrated how the co-evolution between maize and farmers in Mesoamerica resulted in an intimate understanding of crop improvement and the dynamics of diversity. This practical knowledge about crop improvement contrasts with the scientific knowledge of plant breeding, illustrating a difference between use-value and exchange-value. In the second section I connected the milpa to transworld agriculture policies and regulations. I described the neoliberal context in Mexico and its relationship to the global regulatory framework. This context ignores the agroecological contribution of small- and medium-scale farmers who form a majority of farmers in Mexico and are the keepers of maize agrodiversity. In the last section of this chapter I introduced several civil society campaigns in defense of maize and their proposals. These proposals, although sometimes contradictory, demonstrate new ways of framing the maize debates and new strategies for linking radical global political proposals to the everyday practice of growing, exchanging, and eating maize. I described how small- and medium-scale farmers, with civil society organizations, are redefining productivity and biodiversity, and creating new moral, ethical, and political spaces reflecting and contesting the neoliberal landscape. The translocal networks I traced in this chapter provide an example of how food ecologies, economies, and cultures resist, reflect, shape, and are entangled with the story of neoliberalism and food regime formation.

What I emphasized in this chapter are the ways that various ecologies, policies, markets, and social movements overlap. This represents the architecture of the food supply chain and social movements, and the many ways in which the particular intersects with broader processes. The diagrams, coupled with stories of the maize debates in Mexico, served to contextualize the initiatives that follow.

In this discussion of maize networks I explored three tensions: global-local, nature-culture and markets-movements. I illustrated how maize is an agricultural agent; a biocultural assemblage deeply imbricated with the transworld policies that shape trade relations and agrodiversity preservation. At the same time there is a profound cultural politics of maize being played out by civil society and indigenous organizations in Mexico that links local farmers and communities to global social movements. The neoliberal corn regime has transformed the maize economy in Mexico, and in the process galvanized a transnational social movement in defense of maize.

In this chapter, I drew on a rich literature about maize in Mexico. Academic interest in the issue of the maize debates was heightened post-NAFTA and throughout the GM corn controversy. The literature, however, is primarily political-economic in its focus, while the maize story is ripe for a more integrative analysis. By examining the tensions that emerge from the literature, between culture-nature, the global-local, and the markets-movements, I suggest that the notion of biocultural agrifood relations offers a way to examine emerging market relations that contest, and are shaping, the neoliberal food regime.

The initiatives examined in the following chapters are enmeshed within the networks described in this chapter. Although rooted in "place," they connect regionally, nationally, and transnationally to efforts to conserve agrodiversity and promote food sovereignty globally. The three initiatives I describe in the next chapters are reflective of the many diverse strategies being implemented by civil society actors to mitigate the impact of neoliberal policy and free trade. Maize farmers have diversified their production, organized collectively, made contracts with the maize processing industry, and are creating value-added initiatives. The initiatives will serve to illustrate the complexities of maize politics in Mexico and point to the specific practices of biocultural agrifood relations as embedded within and contesting neoliberal globalism.

Notes

1. The work of Lombardi traces in detail the interrelationships between corporate actors involved in various business schemes and scandals. My own diagrams are much less detailed, but try to invoke interconnections among the actors in the maize story—corporate, nonprofit, governmental, entrepreneurial, and vegetal.
2. The Nuestro Maíz project and this example will be elaborated in the following chapter.
3. The Spanish name for INIFAP is Instituto Nacional de Investigaciones Forestales, Agrícolas y Pecuarios. The University of Chapingo is a center for agricultural studies located outside of Mexico City.
4. This is linked to the pro-campesino agricultural policy of the Luis Echeverría administration (1970–1976).
5. Efraím Hernández Xolocotzi (1913–1991) was a preeminent Mexican plant collector, ethnobotanist, and student of indigenous studies who made seminal contributions to the field of agrodiversity in Mexico. He taught at the University of Chapingo. Paczka was a student of Xolocotzi.
6. This section was informed not only by the literature on maize agrodiversity and food security/sovereignty in Mexico, but also by a panel I organized at the Latin American Studies Association 2007 titled Transgenic Corn Debates in Mexico: Biosecurity, Genetic Diversity and Food Security with participants Edit Antal, Yolanda Massieu, Dario Escobar Moreno, Kimberlee Chambers, and Elizabeth Fitting.
7. PROCEDE (Programa de Certificación de Derechos Ejidales y Titulación de Solares, Program for Certification of Rights to Ejido Lands) is the government program that implemented *ejidal* reform in Mexico.
8. Pechlaner and Otero use the term *neoregulation* instead of *deregulation* to acknowledge how the state is implementing new policies such as *ejidal* reform and dismantling programs that support subsistence agriculture, reflecting neoliberal globalism (see 2008, 2).

9. This is also documented in an article I published titled "Local Food Networks and Maize Agrodiversity Conservation: Two Case Studies from Mexico" in *Local Environment* (Baker 2008).

10. The CEC was established in 1994 as a prerequisite for Democratic approval of NAFTA in the U.S. Congress, with a mandate to "facilitate cooperation and public participation to foster conservation, protection and enhancement of the North American environment for the benefit of present and future generations, in the context of increasing trade and social links among our three countries" (CEC 2004).

11. In Spanish these groups are Centro de Encuentros y Diálogos Interculturales, Grupo de Estudios Ambientales, Dirección de Culturas Populares e Indígenas.

12. The Spanish names for these groups are Centro Mexicano de Derecho Ambiental A. C. (CEMDA), Unión de Grupos Ambientales (UGAM), and Estudios Rurales y Asesoría Campesina.

13. Thirteen groups formed this coalition, including Coordinadora Nacional Plan de Ayala, Central Independiente de Obreros Agricolas y Campesinos, and Union Nacional de Organizaciones Regionales Campesinas Autonomas, as well as Greenpeace and the ETC Group.

14. This refers to a UNESCO project to register "intangible cultural heritage," such as oral traditions and expressions, performing arts, social practices, rituals and festive events, knowledge and practices concerning nature and the universe, and traditional craftsmanship. In 2003 Mexico listed El Día de los Muertos (the Day of the Dead) under this convention. See http://www.unesco.org/culture/ich/en/RL/00054.

15. It is interesting that coffee farmers are included in this list of demands. Coffee farmers are well organized, are dependent on exporting their harvest, and receive a low price dictated by the commodity market. Coffee is, of course, not native to Mesoamerica but a colonial transplant. It is also one of the first products in the fair trade movement, a movement that has strong links to other colonial producing countries. This speaks to the importance of global markets to campesinos, and their demands to be fairly compensated within these markets. It also reveals how complex the local/global questions and linkages are.

16. This shift was described by two people during my fieldwork, Silvia Ribeiro of the ETC Group and Aldo Gozalez from UNOSJO (Unión de Organizaciones de la Sierra Juárez de Oaxaca).

3

Nuestro Maíz

THE ARCHITECTURE OF the food networks described in this book goes beyond supply and distribution chains. It includes the political, cultural, and ecological practices that connect farmers and eaters with civil society efforts to create translocally linked niche markets (Fonte 2006; Friedmann 2005a, 261; Petrini 2005). The initiatives that follow represent examples of new, emerging biocultural agrifood relations that re-embed the food economy within democratic social relations and ecological production practices.

The Nuestro Maíz case study illustrates the ways that local food networks and the neoliberal context are mutually constituted, and how these networks draw on new and old forms of organizing markets and movements. The construction of the local food network reflects new relations between place-based markets and translocal social movements. Creative responses to the challenges met by the three initiatives engage new actors in these networks to redefine quality in the food system. The Itanoní Tortilleria case study allows for a deeper understanding of the epistemological roots of biocultural agrifood relations. The Michoacán Centre for Agribusiness demonstrates the role the state can play to nurture local food networks and link dispersed civil society efforts. The interrelationships between particular ecologies, cultures, and economies impact the dynamics of diversity in each distinct context.

The three initiatives provide a concrete way to understand the emergence and meaning of biocultural agrifood relations. Agricultural landscapes and cultural traditions are created and re-created through changing agroecological and sociocultural contexts. The networks described here are translocal and postglobal, rooted historically, self-organizing, complex, and horizontal.

PHOTO 3.1
Tortillas being made at a Nuestro Maíz *tortillería* in Puebla, Mexico.

How do these networks contest and reflect the neoliberal food regime? What are the conditions under which these projects flourish? How do maize movements and networks contribute to our understanding of biocultural agrifood relations? The following chapters shed light on these questions.

Contestation/Reflection: ANEC in Neoliberal Mexico

> We do not and will not accept that we are mere surplus, that we are not productive, competitive, that we are a burden for the country. . . . We will not accept that only large foreign agricultural enterprises have and deserve a place in the countryside, as the federal government has maintained for the last eighteen years. . . . We have been productive in the past; the country has been built through our work. We are productive now despite the unfair competition from imports, and we can be more productive in the future, but only if the policies that favour just a few are abandoned, only if the role of the small and medium peasant farmers is revalued, only if policies toward the countryside are established for everyone, without exclusions or privileges.—From the opening speech at ANEC's Fourth General Assembly, December 8–9, 2000, Mexico City (as quoted in Acuña Rodarte 2003, 139)

This quote, with its focus on the notion of productivity, is a good place to begin a discussion of the three local food networks explored in this book. In this statement, ANEC (Asociación Nacional de Empresas Comercializadoras de Productores del Campo, A.C., or National Association of Campesino Marketing Organizations) is challenging Mexican agricultural and food policy that focuses on shifting into export-oriented fresh fruit and vegetable production at the expense of small- and medium-scale maize farmers. By emphasizing the past, present, and future role of these farmers, ANEC challenges conventional notions of productivity based on production yields and market values. Through its work, ANEC redefines productivity to include the multifunctional agricultural practices historically found in the Mexican countryside, practices that encompass many aspects of social reproduction, what is known in Mexico as *la vida campesina*. This mirrors the proposals of the global peasant movement, Via Campesina, of which ANEC is a member. ANEC's work to "revalue" the role of small and medium peasant farmers implicitly means re-embedding[1] Mexican agriculture and food policy in democratic social relations and ecological production practices, and rebuilding maize supply and distribution chains to include small- and medium-scale farmers. For ANEC this work involves capacity building in the rural sector and transnational advocacy and solidarity work. ANEC develops both political and practical proposals to support their campesino constituency.

In this chapter I discuss ANEC's work in relation to the neoliberal context in Mexico as a way to contextualize the Nuestro Maíz project. I then describe the Nuestro Maíz project and the shortened maize supply chain created through its activities. In the following section, a discussion of tradition and retro-innovation, shifting notions of quality, and issues related to scale and solidarity contributes to a deeper understanding of the theoretical foundations of this book. Finally, I discuss the biocultural agrifood relations forged by ANEC and Nuestro Maíz.

Nuestro Maíz is a project of ANEC, an organization whose membership includes a network of over two hundred producer group cooperatives founded in 1995 to develop marketing opportunities for grain farmers in Mexico. ANEC has two goals. The first is to revalue peasant agriculture, which ANEC views as economically and socially viable, essential for national food security and for the conservation of Mexico's biodiversity. The second is to engage in policy-making processes that affect the lives and livelihoods of Mexico's small- and medium-scale farmers (Acuña Rodarte 2003, 142–143). ANEC is a direct response to the structural changes that took place in Mexico as neoliberal economic policies were implemented in the rural sector.

Over the 1990s, CONASUPO (La Compañía Nacional de Subsistencias Populares [National Company for Popular Subsistance]), the state-run agency that monitored and supported the basic grain market, was completely dismantled. This forced farmers to create their own channels to market their harvest. The enterprises that emerged from this policy shift are known as peasant commercializing enterprises (ECCs). These enterprises addressed social and economic challenges arising from this shift, including mutual support, democratic participation, leadership training, and cooperative management as well as seeking out the best market prices for grain, offering fair interest rates for farmer credit, and programs to increase the productive capacity of small and medium grain farmers (Acuña Rodarte 2003, 139). The ECCs built on a long tradition of cooperative organizing in Mexico's rural sector (see Cornelious and Myhre 1998).

In 1995, as leaders from the ECCs were mobilizing to address shifting agricultural policies, ANEC was formed. ANEC's role is to coordinate the dispersed activities of local and regional ECCs across Mexico, as well as to develop alternative agricultural policy proposals. As stated on ANEC's website,[2] ANEC's mandate is to value and modernize Mexican peasant agriculture, promote and defend food sovereignty, and construct a socially responsible grain industry. This is achieved by providing training, market information, and developing trade networks between local and regional ECCs (Acuna Rodarte 2003, 139–142). In short, ANEC coordinates the marketing strategies and political proposals of regional producer groups.

ANEC's primary economic activity is to market the maize of small- and medium-scale farmers, providing a channel into the national maize market, a market that would not otherwise be accessible to these farmers. By pooling farmers' harvest, ANEC can ensure higher prices, as well as purchase agricultural inputs in bulk, providing a discount to member farmers.

José Armando Joffre Poceros, the director of Nuestro Maíz, describes the history of ANEC in the following way.

> Before 1994 the government began dismantling the grain marketing structures that had existed since the 1940s. In Mexico there was a government agency called CONASUPO that was in charge of marketing maize, bean, wheat, sorghum and rice. By 1994 only two grains were included in the program—maize and beans. When the government started to dismantle CONASUPO, many organizations felt it was necessary to organize collectively to market their grain.
>
> At this time (the early 1990s) the grain market involved many intermediaries—transnational companies, the government, but curiously not farmers themselves. Why? Because a government agency existed that marketed their grain for them. Since 1995, which is when ANEC was founded, collective marketing projects began to be developed. Instead of farmers selling their harvest individually, they organized in small campesino businesses. Slowly, since 1995, the necessary instruments to market their grain have been installed—warehouses, laboratory

> equipment, transportation. ANEC, little by little, started to create rural enterprises that the farmers manage. They are grouped by local markets to sell their maize. (Poceros, pers. comm., February 26, 2006)

The ECCs link local farmers through a cooperative marketing structure developed to provide commercial channels for basic grains including maize, sorghum, beans, wheat, and some other crops from production to processing. For example, the ECCs purchase maize from farmers, pool it with the maize of other producers participating in local and regional ECCs, and sell it on the market when prices are at their highest.

ANEC has taken over key distribution infrastructure once managed by the Mexican government. In the neoliberal context, Mexican agriculture policy has shifted to support export-oriented production (fresh fruits and vegetables), leaving basic grains such as maize vulnerable to market shifts, especially low-priced imported maize from the United States. The national maize market, however, is key to supplying Mexican eaters with white maize for tortillas—a market that represents over 40 percent of maize trade in Mexico (CNIPMT 2008). In the post-NAFTA process of shifting agriculture policy programs, ANEC has provided farmers with a channel to market their grain.

A substantial infrastructure is required to channel grain to the national market, and through ANEC's political connections and lobbying efforts, this infrastructure has grown. Approximately 180,000 small- and medium-scale grain farmers are members of over 220 ECCs in twenty Mexican states. In recent years, six hundred thousand tons of maize was collected and traded through 135 grain warehouses. In total, ANEC trades approximately 10 percent of Mexico's production of basic grains, earning farmers a price premium of 10–15 percent on top of the national market price for maize (Acuña Rodarte 2003, 141). This activity is dependent on suitable storage and warehouse facilities. The infrastructure has been built over time, with 110 collection centers acquired from the federal government when CONASUPO was dismantled.

An umbrella enterprise, SIACOMEX (Integrated Storage and Commercialization System of Mexico), has been created by ANEC to coordinate grain collection and trade at the national level and to facilitate access to larger markets. Leadership training, commercial support, market information, and technical assistance are provided along with grain marketing. ANEC's farmers have created a market for their agricultural products and have developed some innovative programs to support farmers across Mexico. ANEC's marketing services are primarily funded by the government program Alianza Para el Campo (Alliance for the Countryside) with loans to build the necessary infrastructure and grants for capacity building and technical assistance.[3] The reliance on government funding shapes the direction of the organization's initiatives. For example, government funding for rural enterprise development resulted in Nuestro Maíz.

The Mexican government dismantled their key program (CONASUPO) supporting small-scale farmers, ensuring that the services were continued by allowing ANEC to take over the EECs. In a neoliberal marketplace, this is a familiar state strategy—divesting of programs while civil society organizations struggle to provide essential services. The Alianza para el Campo program consolidates this trend by providing loans and funds to a number of nongovernmental actors (primarily private sector, but also nonprofit) in the agricultural sector to invest in agricultural technologies and improve productivity.

This short history of ANEC illustrates how neoliberal policy has resulted in new economic strategies from the civil society sector. Even though key government programs, such as CONASUPO, were dismantled, the services offered carry on, taken up by the civil society actors working on policy alternatives and struggling to support their constituency—in the case of ANEC, Mexico's small- and medium-scale farmers. ANEC's strategy draws on Mexico's strong tradition of cooperative organization (Cornelious and Myhre 1998) to create networked enterprises that support their campesino constituents.

Civil society organizations are creating infrastructure to strengthen regional and national food supply chains, and the activities of ANEC are an example of what Bryant and Goodman (2004) call "entrepreneurial developmentalism." In the case of ANEC, dispersed, self-organizing groups of farmers from around Mexico joined together through horizontal networks to collectively market their grain and generate policy proposals in the face of NAFTA. These efforts were historically rooted in the programs dismantled under neoliberal policy, and are examples of the "active work of preservation" described by Alvarez (2005). These efforts, although rooted historically, are characterized by innovation and translocal linkages, as will be described in following sections. A close examination of the Nuestro Maíz project illustrates how maize "*adds something* to the story of neoliberalism" (Braun 2006), suggesting that maize's agency complicates the political and economic goals of the project.

Valerie Tarasuk and J. Eakin (2003) and Graham Riches (1999) examine the role civil society plays in providing emergency food distribution in Canada, providing examples of how alternatives such as food banks and community food projects flourish in the face of neoliberal policy shifts. As state responsibility is downloaded to the charitable or nonprofit sector, the approach to service delivery shifts. For example, ANEC's program to market the maize of small- and medium-scale farmers in Mexico is based on building the capacity of producer groups to manage their own organizations; participate in broader regional, national, and transnational movements; and contribute to the decision-making processes of the organization. The CONASUPO program had no such focus, and Joffre Poceros acknowledges the challenges related to taking this approach in a context where farmers were left out of decision making historically and thus lack experience in democratic practice. This reflects the long history of clientelism in Mexico, rooted in the paternalistic role of the state, a legacy many civil society groups struggle with. ANEC's work engages farmers in a democratic decision-making process that includes annual regional and national meetings where farmers can shape the organization's policy and programs.

ANEC's work is embedded in the transworld agriculture policies of the World Trade Organization and the General Agreement on Tariffs and Trade, and exemplified by NAFTA. Their activities are driven and shaped by neoliberal trade agreements and policy processes, as well as the increasing interest in shortened food supply chains by civil society actors in the face of shifting government involvement in the food and agriculture sector. Although ANEC's work is place-based, through the local producer groups, it is not place-bound but linked regionally and nationally to other producer groups supported by ANEC and transnationally through the campaigns that ANEC is active in globally (discussed later in this chapter). Some of these linkages are created through the maize itself, as it circulates locally and nationally through communities, grain warehouses, processing plants, and kitchen tables. Other linkages are created through capacity-building efforts that bring farmers together from different communities across Mexico. This sense of circulation and intersection transcends and de-purifies the local (Gupta 1998), drawing attention to the global nature (Braun 2006) of efforts such as ANEC's Nuestro Maíz.

ANEC's project is one of economic experimentation within the context of neoliberal policies. The practical work of selling maize is coupled with oppositional politics aimed at advocating for a policy environment more favorable to small and medium farmers. These are the economic and political conditions under which ANEC's programs are flourishing. The organization fuses maize markets and social movements within a translocal context that stretches beyond Mexico. In this sense, ANEC is "performing the economy otherwise," expanding capitalist notions of productivity to include broader livelihood strategies of Mexican campesinos. For ANEC, shifting from a focus on market-based production to social reproduction means acknowledging the links between local ecologies and culture through a focus on sustainable campesino livelihoods. Looking in more detail at one of ANEC's newest economic strategies, it is possible to examine the biocultural agrifood relations of the organization. The following section introduces the Nuestro Maíz project.

Linking Farmers to Eaters: The Shortened Maize Supply Chain

The Nuestro Maíz project links maize farmers to eaters through a shortened supply chain. Maize grown by local farmers is sold to a producer cooperative. Some of this is funneled to the Nuestro Maíz project's *nixtamal* plant. The masa is then distributed to neighborhood *tortillerías*. The *tortillerías*, staffed by farmer family members, provide additional income to enable campesinos to stay on the land and continue producing maize. The following section describes this chain, focusing on the biocultural agrifood relations of the project and drawing out the concepts of social entrepreneurship, tradition, innovation, and retro-innovation in the context of their work.

The Nuestro Maíz project—adding value to the maize of campesinos and developing local markets for this maize through tortilla chains—was a new strategy for ANEC, deviating from their past work to channel the maize produced by small- and medium-scale farmers into the national market. As maize prices fell in Mexico from the mid-1990s, ANEC sought new strategies to make maize production in Mexico part of a sustainable rural livelihood. For ANEC, Nuestro Maíz is a rural enterprise development project that addresses a number of social and economic issues facing rural Mexico. The problems addressed by Nuestro Maíz include lack of employment, low maize prices, and poor quality tortillas. ANEC's social entrepreneurial efforts address environmental issues by promoting agroecology, seek ways to diversify economic opportunities available to small- and medium-scale farmers, and seek synergies along the maize production-consumption chain. In summary, Nuestro Maíz is a value-added rural development project created to complement ANEC's work to support small- and medium-scale farmers in Mexico, the first value-added activity ANEC has undertaken.

The Nuestro Maíz project's goal is to create regional value-added markets for the maize of participating producer groups. The project uses locally grown maize to make tortillas sold in surrounding communities. The project has seventeen *nixtamal* plants in nine Mexican states with a network of over two hundred *tortillerías*. Six hundred workers are employed by the project (in either the *nixtamal* plants or the *tortillerías*), 70 percent of whom are family members of the producer groups. "It is a circle involving campesino organizations, a commercial project, the generation of added value for maize and the creation of employment, all in the same community" (Joffre Poceros, pers. comm., February 26, 2006).[4]

Each producer group grows maize that is sold to a central warehouse where it is processed and funneled into the national maize market. Through the Nuestro Maíz project, however, a small portion (currently less than 10 percent) of this maize is diverted to *nixtamal* plants, adding value to the participating farmer's maize. For example, a kilo of maize is worth 1.5 pesos on the market. If the same kilo of maize is made into tortillas it can be sold for 8.5 pesos.[5] Instead of selling the maize as grain to the national market, it is sold as tortillas, increasing the value of maize for farmers. The long-term goal of the project is to add value to 50 percent of the farmer's maize through the Nuestro Maíz project.

This case study focuses on the activities of the San Antonio Atotonilco producer group in Puebla (see map 3.1). This area of Puebla is not irrigated, and so campesinos depend on the seasonal rains for their maize, beans, and squash crops. The San Antonio Atotonilco producer group is well organized, having collectively marketed their grain and participated in campesino movements in Mexico before ANEC's existence.

All of the San Antonio Atotonilco farmers grow maize as well as beans, squash, and other crops. The thirty-eight farmers who are part of this group each have anywhere from 3 to 10 hectares under production. The San Antonio Atotonilco producer group has a total of 175 hectares under maize production and 200 hectares producing other crops. Luis Ramero is the president of the producer group. He and his extended family farm 40 hectares. The land is fragmented (not contiguous parcels). Luis himself has 2 hectares, his father has 10, and the rest is rented from other farmers. Of the 40 hectares, 30–35 hectares is under maize production; the rest is other crops, including squash

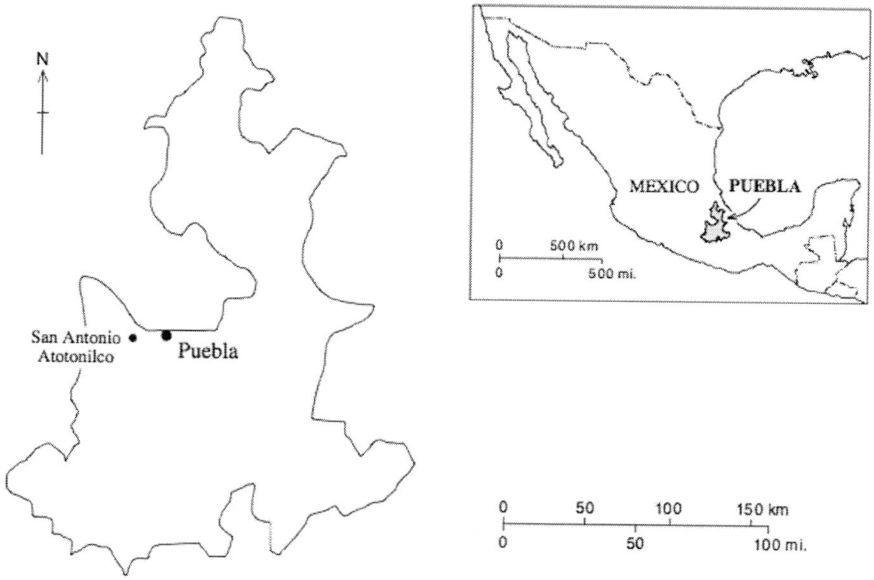

MAP 3.1
San Antonio Atotonilco, Puebla. Map by Carolyn King.

and beans in rotation. None of the land is irrigated. The farmers' production methods are conventional. Agricultural inputs such as chemical fertilizers, pesticides, and herbicides are used by the farmers. ANEC is questioning conventional agricultural methods and is working with farmers to shift to agroecological production methods. Participating in the producer group means access to financing and bulk prices for production inputs. Through the association with ANEC, the farmers have opportunities for capacity building, networking with other producer groups across the country, and engaging in food sovereignty work nationally and internationally.

The maize grown by these farmers is sold to the grain warehouse, some of which is diverted to the Nuestro Maíz *nixtamal* plant. The *nixtamal* plant transforms the maize into masa, which is delivered daily to thirty *tortillerías*, established by the Nuestro Maíz project and located within a twenty kilometer radius of the plant in the surrounding towns and villages. The tortillas are freshly made at each *tortillería* and sold to the local community. Members of the farmers' families and local community work at the *tortillerías*. Customers are community members. Occasionally institutional customers such as hospitals or schools are sought as well.

The *nixtamal* plant, in San Antonio Atotonilco, Puebla, was built in 2003. The project is still paying off the construction debt, but the farmers will eventually reap the profits of the project. The San Antonio Atotonilco producer group owns the local technology and shares the profits and risk of the business among the members, operating like a cooperative.

PHOTO 3.2
The industrial-scale *nixtamal* technology.

The industrial-scale *nixtamal* technology, shown in photo 3.2, was common in Mexico throughout the 1960s and 1970s, but this *nixtamal* technology and the resulting masa tortillas have faced competition from the corn flour tortillas of MASECA/Gruma.[6] The large soaking vat and grinder can handle a capacity of two thousand kilograms of maize each day, but in San Antonio Atotonilco the project is operating at 50 percent capacity. This is because there have been some start-up challenges, limited demand for the tortillas, and production challenges related to maize quality (described in following sections).

Each evening the maize is put into the vat to soak with *cal* (an alkaline lime solution). The maize and *cal* are heated briefly, then the temperature is lowered and the mixture is soaked overnight for twelve hours. Early the following morning the maize is washed and rinsed before being ground. The powerful grinder, made from Mexican volcanic rock, can produce fifty kilograms of masa in five minutes. Two employees run the plant and work from 4:30 a.m. to noon.

The masa is delivered to the thirty *tortillerías* by motorbike and truck by 11:30 a.m. Staff at the *tortillerías* take the masa and begin making tortillas in preparation for the mid-day meal rush. The *tortillerías* have mechanized tortilla presses, so up to a thousand tortillas can be made in an hour. The tortillas are sold for 8.5 pesos/dozen.

Marketing materials describe the tortillas as made from 100 percent Mexican maize, preservative free, and free of glues commonly used to bind flour tortillas (see the tortilla packaging in photo 3.3).

PHOTO 3.3
Nuestro Maíz tortillas and labels.

The Nuestro Maíz supply chain diverts maize that would otherwise be funneled into the national maize market to create direct linkages between local farmers and eaters. Nuestro Maíz reflects and contests neoliberal policy processes, illustrating how the global permeates the local. The project is a result of ANEC developing programs that respond to the government funding available, which in turn reflects the neoliberal trend to shift agricultural subsidies away from the subsistence maize agricultural economy and towards rural enterprise development. The project is created from the ecological and cultural traditions of campesino maize production and invokes a cultural politics of place through its marketing of 100 percent Mexican maize. The cultural and culinary importance of maize is used as a marketing tool by the project, and production is geared to support small- and medium-scale campesino livelihoods. Based within a broader organization that connects regional farmers nationally through alternative policy proposals, as well as transnationally through global food sovereignty campaigns, Nuestro Maíz provides an example of how social movements and markets exist in dynamic tension. Illustrative of biocultural agrifood relations, the Nuestro Maíz project is creating new networks and relationships between markets and social movements within a translocal context. In the process, interesting contradictions and opportunities for creating local food networks are revealed.

The Qualities of Maize, Quality Tortillas

Nuestro Maíz can be characterized as an *empresa social* (social business), as defined by Maria Dolores Villagomez (1999, 215). *Empresas sociales* "are economic units managed by poor people with the goal of maximizing the capital invested in order to generate some social value" (215). Financial, organizational and social, and political aspects of these businesses include production that responds to a social need, workers involved in planning processes, capacity building for members, the creation of an economic alternative, and building social and political networks. Also articulated by Villagomez is the accounting for the cultural, ecological, and social costs of production (216).

The work of Nuestro Maíz sheds light on shifting notions of tradition, innovation, and retro-innovation, as explored by Kevin Morgan, Terry Marsden, and Jonathan Murdoch (2006). It is "often those regions that have traditionally been regarded as 'marginal' [that] now begin to display the most conducive conditions for the development of alternative agrifood networks and new forms of value-capture" (86). Morgan et al. examine how these social and ecological entrepreneurial efforts in local food networks form new production-consumption linkages. Similar to the examples from California,

Tuscany, and Wales described by Morgan et al., Nuestro Maíz draws on the historical importance of maize, the strong tortilla economy, and existing *nixtamal* technology.

Using the *nixtamal* technology goes against the industrial practice of using maize flour in the tortilla industry. The adaptation of the *nixtamal* technology for a rural development project is an example of a "retro-innovation" or "backward-forward technological innovation" (Morgan et al. 2006, 87). The Nuestro Maíz project reclaimed a technology that was largely left behind by the conventional tortilla sector, in order to distinguish itself from the *harina* (flour) tortilla industry. By using the *nixtamal* technology, the project aims to appeal to a nostalgic, concerned consumer who relates tortilla quality with traditional methods, validating the commonly made connection between Mexican culinary traditions, food quality, and nutrition (Pilcher 2005; Vizcarra Bordi 2006).

The niche market for *nixtamal* tortillas is an example of how "tradition" is historically constructed and reconstructed, and how production and consumption practices are constantly being remade through intersecting global and local processes. Industrial tortilla production, since the mid-1970s, has replaced masa dough with ground corn flour, which is nutritionally inferior (Rivera et al. 2004). The informal market has responded to the Mexican people's ongoing appreciation for maize products made with masa. *Tortilleras* (women selling tortillas and other corn snacks) can be found in all markets and sometimes outside of supermarkets, offering "homemade" tortillas. As people's consumption practices become integrated into the global economy and they shop more frequently at supermarket chains (see Chavez 2002), the purchase of tortillas made from corn flour has increased. In the competitive Mexican tortilla market, Nuestro Maíz uses a number of strategies to distinguish their product. They focus on branding the tortillas with an emphasis on quality. They are also expanding their products and will be opening a plant that produces snacks and maize sweets, in order to diversify sales.

When describing the process of creating Nuestro Maíz, the project director, Joffre Poceros stated,

> We asked ourselves, "What do we need to do to start selling tortillas?" The market for tortillas in Mexico is already saturated. We don't need more tortillerías. But tortillerías in Mexico are unhygienic, disorganized, and lack standards. We thought about developing a tortillería model that addressed these issues. The Nuestro Maíz model processes the maize of our associates; uses the pre-hispanic process of maize, water and calcium; does not use glues, additives or preservatives. And it is something that the regional organizations can administer. (pers. comm., February 26, 2006)

For Joffre Poceros, notions of quality are centered on cleanliness, hygiene, and a preservative-free, "pure" product. Nuestro Maíz markets these qualities, appealing to the consumer's desire for a hygienic tortilla-making process by focusing on its high processing standards. Quality also refers to the quality of the maize. Joffre Poceros describes the maize as GMO (genetically modified organism) free, clean, sorted, and rinsed (but not organic). The *nixtamal* process gives the tortillas an "authentic" taste and higher nutritional value, distinguishing Nuestro Maíz's product from the "contaminated" (both from GMOs and unhygenic production practices) products available elsewhere.

The flyer in figure 3.1 is a marketing tool for Nuestro Maíz, articulating the quality of their tortillas. The flyer states, "Do you know what maize your tortillas are made from? Our tortilla is FREE . . . FREE of transgenic maize, FREE of conservatives, FREE of maize flour. Nuestro Maíz—100% Mexican maize. Tortillas made from the traditional process: maize—*masa*—tortilla."

FIGURE 3.1
An advertisement for Nuestro Maíz tortillas.

Eaters, however, have challenged how the project has constructed the qualities of the Nuestro Maíz tortillas, articulating different notions of quality. Customers have complained about the color and texture of the tortillas. The tortillas the Nuestro Maíz eaters are accustomed to are a particular color (white) and a particular texture (they hold together under a moist *guiso*, or filling). Eaters have also complained that the tortillas vary in quality from day to day and week to week. The heterogeneous characteristics of the tortillas, as they reflect the particular batch of local maize, clash with the homogenizing tendencies of consumer demands. These demands and consumer expectations are shaped by global production processes, illustrating the penetration of monocultural agroindustrial practices. Tastes have changed through exposure to mass-produced flour tortillas.

This example illustrates how quality has different meanings for different actors in the food chain (Mansfield 2003). Beyond quality debates forged at the negotiation table of the Codex Alimentarius, there are parallel and divergent debates forged in local food networks. Joffre Poceros draws on the discourse of quality related to production standards, while eaters express notions of quality based on expectations of a homogenous product and the flavor and texture of mass-produced tortillas.

The *nixtamal* process has presented other challenges for Nuestro Maíz. The same maize landraces that the project wants to conserve have proven difficult to adapt to the *nixtamal* technology the project uses. The criollo maize did not adapt well to the larger scale *nixtamal* processing technology. The criollo maize flour content[7] varies from landrace to landrace, making it difficult to standardize the *nixtamal* process. "When we began this project we thought we had the productive conditions to be successful. But we didn't" (Jorge Aguirre Alonso, pers. comm., February 25, 2006). On a tour of the *nixtamal* plant a worker commented, "We thought it [the *nixtamal* process] would be very standardized and we would soak the maize for a certain amount of time. Instead we need to rely on our senses. The maize tells us through smell and taste if it has soaked for long enough" (pers. comm., February 27, 2006). Criollo maize thus also shapes the industrial and commercial process.

The heterogeneity of criollo maize and criollo maize products is an example of "the persistence of the organic" (Goodman and Watts 1997), as well as how commodification is resisted by nature (Braun 2006). In this sense, criollo maize expresses agency within the maize supply chain, challenging social entrepreneurial efforts to build markets based on static notions of tradition and diversity. These conflicting notions of quality, coupled with the challenges presented by the criollo maize, have resulted in a rigorous crop improvement program involving local farmers and national researchers.

The Nuestro Maíz crop improvement program works closely with agronomists and crop improvement specialists from INIFAP (the National Institute for Forestry, Agricultural and Livestock Research) to improve the maize landraces used by local farmers in an effort to make the criollo maize meet the technical needs of the *nixtamal* plant as well as consumer demands for a uniform tortilla. The San Antonio Atotonilco farmers, in order to address the heterogeneity of their maize, are planting hybrid varieties of corn beside their maize landraces to improve the uniformity of the grain. The twinning of rural development goals (improving the economic viability of maize production) and agrodiversity conservation has proven difficult for Nuestro Maíz, and the project reveals the tension between the two.

Nuestro Maíz works with farmers to improve their farming systems, with a focus on maize crop improvement and the introduction of suitable hybrids through work with extension workers and crop improvement specialists from INIFAP. Agricultural extension, fundraising, administrative support, and maize improvement initiatives are facilitated by the central Nuestro Maíz organization located in Mexico City and funded by a grant from Alianza para el Campo.

> The model surfaced from the threats to and the importance of criollo maize in Mexico. When criollo maize is used by campesino families to make tortillas—a few each day—it functions very well. But when you take this maize to an industrial process, it doesn't function in the same way. We created the industry, the infrastructure, and then realized that the primary material was not adequate. But we already had the enterprise installed, we already had the *tortillerías*, the plant, the vehicles, everything. The criollo maize functions very well when a campesino woman uses it to make tortillas daily for her family. But when it is processed in large quantities, it doesn't function the same way. So we had to find a way to resolve the issue of the maize quality. We are working with INIFAP to improve the varieties of maize grown in each region. We are improving the criollos so they can be used in the masa and tortilla industry. But there are places where we've had to change the criollos, and so what we are doing is producing hybrid maize ourselves, for our *nixtamal* plants. (Poceros, pers. comm., February 26, 2006)

Joffre Poceros describes the rationale for the crop improvement program, revealing challenges related to scale. As the use of criollo maize shifts from the subsistence activities of campesino families to the market-based activities of Nuestro Maíz, the quality requirements shift. Referring to the same challenges, the president of the Atotonilco producer group describes the way that maize production practices have been shaped by the project.

> When we started with the tortilla production we were using every type of maize. There was a tremendous mix, and we couldn't control the process because we didn't have quality grain. Within two years we have been able to produce grain specifically for masa/tortilla production, and separate it from grain for forage. (Arturo Jimenez, pers. comm., February 27, 2006)

In these quotes the tension between criollo, improved, and hybrid varieties is evident. A project that aims to conserve criollo maize is introducing hybrids to farmers, a seeming contradiction. Yolanda Salinas Moreno distinguishes between the quality of maize appropriate for self-sufficiency and the quality of maize needed for the market and elaborates the goals of the crop improvement program.

> The criollo varieties in this region [Puebla] are very soft. They don't translate into the kinds of tortillas eaters are used to—a tortilla with a bit of resistance that can hold the filling's liquid. Hybrid varieties have this resistant quality. In our program with Nuestro Maíz we evaluated the maize of the farmers. What quality was it? What qualities did it need to have to meet the demands of the urban market? We introduced hybrid varieties that would address these quality issues and gave the farmers access to this improved material, developing the capacity of farmers to grow the seed themselves. Now they can have access to improved seed cheaply. This is another thing about the market. The prices for improved seed are inaccessible for small farmers. This is the role of institutions like INIFAP. We can provide a social benefit to small farmers. (pers. comm., February 24, 2006)

The issues raised by Nuestro Maíz's crop improvement program illustrate the internal contradictions faced by local food networks. These contradictions challenge notions of purity related to agrodiversity and criollo maize. In many respects this is an example of how market forces have shaped, and continue to shape, agrodiversity. The farmers participating in the Nuestro Maíz project have not been isolated and operating outside the marketplace. Their production practices reflect their position straddling self-sufficiency and the market. The networks created by the project—linkages to the local market through a value-added project that connects their maize directly to eaters, as well as the institutional relationships of the Nuestro Maíz project, such as the relationship between ANEC and INIFAP—shape production practices.

Through their involvement in the Nuestro Maíz project these farmers are now "improving" their criollo varieties thorough the introduction of hybrid maize varieties. They have changed their seed saving practices and now select seed in the field instead of after harvest to ensure they choose maize that is shorter and sturdier. Participating farmers grow and distribute their own improved varieties. The farmers are incorporating new skills, building on their historical practices of seed saving. Again, the tensions between traditional and new practices are apparent. Some farmers are resistant to these changes, as Salinas Moreno notes:

> It is very difficult to convince the farmers to change their production practices. To convince farmers we established a trial of the improved varieties in the plots of some of the farmers. Last year it was successful. It was a very difficult year

> because it didn't rain. The weather was very erratic. Some of the improved varieties, despite these poor conditions, were more productive than the farmer's criollo varieties. We can't convince farmers by telling them to grow something. They do not have the confidence that these varieties will do well. Historically, hybrids required technologies and inputs the farmers didn't have access to: fertilizer, irrigation. We have developed hybrids that don't need this, that farmers can produce themselves. (pers. comm., February 24, 2006)

This approach to hybrid development focused on the conditions of farmers in a particular region differs from hybrid development focused on yield and corporate profit. A farmer articulates the conflict between traditional practices and the profitability of the Nuestro Maíz project.

> We have done a lot of work in the area of crop improvement. This has focused on raising the consciousness of the farmers, and changing their ideas based on long traditions of growing criollo maize. It has been very difficult. Another challenge that we face is changing the eaters' ideas that the best tortilla isn't a white one. This is a new concept for most. This is key to making the plant profitable for us. (Ramero, pers. comm., February 25, 2006)

In this quote it is suggested that farming traditions and practices need to be shifted to adapt to the requirements for a successful Nuestro Maíz project. Particular conceptions of criollo maize are invoked, as well as particular notions of quality related to tortillas. Amado Ramirez Leyva, founder of the Itanoní Tortillería described in the following chapter, helped me make sense of these contradictions and challenges. He described two overlapping worlds of diversity.

One of these worlds represents the diversity of criollo maize, which can be categorized by "races" or thought of as an infinite expression of genetic diversity based on farmer practices, cultures, ecologies, and economies. This is a deep diversity, connected to agricultural practices and reflecting historical ties to particular people and places. The second world of diversity represents the static diversity of improved and hybrid maize. This world includes a spectrum of crop improvement and breeding activities, but is based in scientific notions of genetic resources—improved criollos, improved varieties, synthetic varieties, hybrids, and genetically modified (GM) corn. The crop improvement program at Nuestro Maíz operates in the space where these two worlds overlap. Criollo maize is improved through the introduction of hybrids, a partnership between farmers and scientists, in situ, or in the fields, building on the farmer's agricultural practices. This partnership between farmers and scientists is a key aspect of biocultural agrifood relations and the democratization of quality, place-based food networks. Harriet Friedmann labels this partnership "scientist-farmers" and "farmer-scientists" to explain how the skills and knowledge of both can come together creatively to address challenges and tensions that emerge in food networks (Friedmann 2003).

The issues associated with heterogeneity and quality have galvanized a whole new focus for Nuestro Maíz, extending the food network to include crop improvement and building local capacity in this area. The crop improvement program was an unanticipated consequence of the Nuestro Maíz project, engaging crop improvement specialists in the project. In the Nuestro Maíz example, local production practices and maize diversity is being shaped by the technological requirements of the *nixtamal* plant, as well as by consumer preferences influenced by global production processes and markets. However, maize can be seen to retain agency in the way that it is galvanizing crop improvement processes, bringing together the two worlds of diversity just described. Despite these interesting contradictions, maize continues to actively shape the project, as it has shaped agricultural landscapes and food economies historically. Maize retains agency in this dialectical relationship between campesino agricultural practices, the industrial demands of the *nixtamal* technology, consumer demands for uniform tortillas, and the cultural importance of maize based on certain ideas of authenticity.

The notion of biocultural agrifood relations helps to make sense of the crop improvement work undertaken by Nuestro Maíz in partnership with INIFAP. The program enlists farmers and seed scientists in a democratic process to extend the food network and improve maize for the local market. This is an example of how hybridization can be democratic when it involves farmers in the process, as illustrated by Kloppenburg (2004) and discussed by Esteva (1996) in an article on clashing knowledge systems related to agronomy and plant science. Both authors describe how seed improvement for corporate profit has displaced more democratic forms of crop improvement.

This section illustrates how the nature-culture, global-local, and markets-movements tensions are deconstructed and reconstructed through biocultural agrifood relations. The worlds of diversity are neither static nor place-bound. On the ground, in the milpa, these tensions are constantly in play, a dynamic illustration of how agrodiversity interacts with specific people, institutions, technologies, and policies. Criollo maize economies and cultures reflect these interactions and the dynamics of diversity ebb and flow accordingly. The cultural politics of maize is used by the Nuestro Maíz project and results in contradictions and challenges when multiple goals (such as rural development and agrodiversity conservation) are pursued.

Scaling Up and Out through Building Solidarity

Nuestro Maíz is experimenting with a rural development strategy to keep local maize and maize profits circulating in communities across Mexico. Working with Mexico's campesino farmers who have historically been politically strong and nationally supported, the project hopes that these efforts can be replicated and adapted across regional and community contexts. The national scope and political connections of ANEC facilitate this scale of organizing and the rural development process. For example, ANEC has been able to access funds from the government and bring farmers from different regions together to discuss their future.

The success of the San Antonio Atotonilco Nuestro Maíz project is a result of the strength of the producer group. The Nuestro Maíz model has been difficult to replicate across the country and the success of each individual project has depended on the participating producer group. Many producer group projects, despite initial capital investment in the *nixtamal* plant, have not yet begun operating, or have not been able to make their loan payments. There are many factors that contribute to the economic success of a project such as Nuestro Maíz, but the people interviewed described the successes and failures of the various Nuestro Maíz projects as related to the capacity of the producer groups and their abilities to take on large entrepreneurial, capital intensive projects.

ANEC understands the need for producer-group capacity building and its challenges in a context where the price of maize is fluctuating wildly and government support for campesinos is dwindling. Workshops are held for farmers to train them in the skills needed to market their grain, work cooperatively, understand the global market, and use agroecological production methods. This capacity building serves the farmers, but also the larger organization as it responds to policy shifts at the federal level.

The elimination of the remaining barriers to importing maize to Mexico in January 2008 under NAFTA further undermined the success of the Nuestro Maíz project. Campesino farmers expressed concern about their ability to continue growing maize in the future and even about staying on their land. In this context, it is questionable whether capacity building will be able to address larger structural issues faced by Mexican farmers.

As stated by Victor Suarez, ANEC's executive director,[8] ANEC's success has been to "value the national capacity that we have as a country to produce healthy food in adequate supplies for everyone who lives in the country, and this speaks to the revaluation of indigenous peasant farmers, as well as sustainable, organic, and other alternative forms of production" (quoted in Acuña Rodarte 2003, 144). This quote illustrates the tension between the ability of Mexican campesinos to adapt to shifting policies and their marginalization in the current policy context.

In its efforts to create alternative forms of production and distribution, ANEC's Nuestro Maíz project has faced some challenges. Despite funding for capacity building, it struggled to support producer groups in disparate areas of the country while it undertook a large, capital intensive project. This left critical work in the hands of regional producer groups who did not have the capacity to successfully implement the project, leaving them in debt and conflict. Furthermore, the technical challenges were not anticipated and resulted in a whole new aspect of the project focused on crop improvement that is difficult and expensive to replicate across so many regional contexts.

Another issue has arisen related to marketing the tortillas produced at the Nuestro Maíz plants. At the Nuestro Maíz plant operating within Mexico City there have been protests by other local tortilla farmers. The neighboring *tortillerías* have complained that it is unfair for a (partially) subsidized tortilla operation to come into the neighborhood and take over their business. This represents an interesting dilemma for local food networks that are not-for-profit or state subsidized. In the case of this particular Nuestro Maíz project, an agreement was made between ANEC and the local *tortillerías*. The Mexico City Nuestro Maíz project only sells to institutional clients—hospitals, government buildings, schools, and so on. In the case of Nuestro Maíz, the extended networks—linkages with government agencies, crop improvement laboratories, and marketing expertise—became a source of conflict with other masa tortilla vendors in the local market.

The example of the protests against Nuestro Maíz in Mexico City illustrates that as local food networks scale out (Johnston and Baker 2005) to more regions, local market differences need to be considered and addressed. This is the challenge for emerging translocal markets of niches reflecting biocultural agrifood relations. As these local networks extend and galvanize public support, opportunities for shaping policies—at municipal, state, and global levels—increase. In Brazil, for example, the state has committed to replicating the food security programs developed in Belo Horizonte, but this program of expansion and replication has faced difficulties (Rocha 2001). The Belo Horizonte experience reflects the importance of democratic process as well as specific community contexts and illustrates the tension between the market and social movements as models for biocultural agrifood relations are translated across regions from one place to another. The local context still matters, and it is difficult to parachute into a community with cookie-cutter programs.

Based on the experience of the seventeen producer groups who have started Nuestro Maíz projects, it is questionable whether the productive capacity and conditions exist to replicate these initiatives across Mexico. In many ways, the

Nuestro Maíz project is highly centralized, with the project concept originating in the head office of ANEC in Mexico City, an organization without any previous experience in the highly competitive tortilla sector. As the Nuestro Maíz project was implemented across the country, the head office struggled to support the regional producer groups. The historical strength of each producer group—the ability to work together on a complex, multifaceted project—directly impacted the success of the projects.

As part of ANEC's capacity building and political work, the organization is involved in local, national, and transnational advocacy, solidarity, and policy making. ANEC channels substantial resources into advocacy and solidarity related to agriculture and food policy in Mexico and beyond. One example of this is the development of ANEC solidarity markets. These efforts support regions in Mexico that are experiencing production shortfalls due to drought or other environmental or social factors. Grain is supplied directly to those regions from ECCs with a surplus to ensure regional food security.

This solidarity work expands from across Mexico into the international realm with affiliations and collaboration with national and international farmer and rural organizations including Mexican coffee producer and rural credit organizations, producer groups across the Americas, and transnational organizations such as Via Campesina, the InterAmerican Agriculture and Democracy Network, and the Global Peasant Agriculture and Globalization Network (Acuña Rodarte 2003, 142). This work contributes to the international policy platforms of these organizations and coalitions that critique neoliberalism and call for food sovereignty.

Compounding Mexican farmers' struggle for economically viable production schemes was the GM corn issue. ANEC has been involved in the GM corn issue, supporting and working in partnership with the anti-GM corn efforts of groups such as Greenpeace, ETC Group, and the coalition In Defense of Maize, articulating the important role its farmers play in preserving maize agrodiversity. Hence, the Nuestro Maíz project links issues of rural development and agricultural biodiversity not only by attempting to make maize production more economically viable, but also by linking farmers to each other across the country and to farmer movements across the globe. For example, San Antonio Atotonilco farmers have participated in demonstrations against NAFTA and in the El Campo No Aguanta Más campaign and in national producer meetings held annually by ANEC.

The linkages that ANEC has with political and civil society actors are strong. ANEC is a lead organization in many efforts related to defending maize. For example, on the eve of the complete dismantling of tariffs under NAFTA, ANEC was one of the lead organizations behind a renewed Sin Maíz, No Hay Pais[9] campaign that asked Mexicans to "put Mexico in your mouth," referring to Mexican maize consumption. The signatories of this declaration include national producer groups, regional or state producer groups and international nongovernmental organizations. The demands include renegotiating NAFTA, banning GM corn, implementing policies that support small- and medium-scale farmers and urban eaters, and protecting the livelihoods and cultures of rural Mexico. Nuestro Maíz is an example of the diversity of strategies used by groups such as ANEC in the face of agricultural and dietary transition.

ANEC has had some success in influencing Mexican agricultural and rural policy decisions, although the overall direction of government policy has not been shifted. Most significantly, ANEC convinced the government to charge tariffs on above-quota imports of corn, beans, and sorghum from 2000 to 2006 and defeated a Fox administration motion to decrease the agriculture budget, winning an increase of more than $400 million pesos (Acuña Rodarte 2003, 144). Perhaps most important are the numerous policy documents and proposals written and put forward by ANEC to influence agricultural and rural policy in Mexico. The internal challenges faced by ANEC as an underfunded, nonprofit organization working to support small- and medium-scale farmers in Mexico are dwarfed by the formidable challenge of renegotiating NAFTA, addressing the dumping of U.S. grain on international markets, and shifting Mexican agricultural and rural policy.

The Biocultural Agrifood Relations of Nuestro Maíz

The notion of biocultural agrifood relations offers an analytic tool for understanding the challenges faced by Nuestro Maíz and ANEC. Replicating models across diverse ecological and cultural contexts does not take into consideration regional differences, the qualities of criollo maize in each place, and the historical capacity of farmers to organize and engage in complex entrepreneurial projects. The tensions followed throughout this book—nature-culture, global-local, and markets-movements—create friction for these transplanted initiatives. Perhaps this friction will result in new creative solutions (Tsing 2005), but the neoliberal political context makes it difficult for local food networks to flourish. Their success is related to particular ecological, cultural, and economic conditions. In San Antonio Atotonilco, Puebla, the conditions allowed for the Nuestro Maíz project to begin operation, but not without unanticipated challenges.

Rural farmers who are involved with organizations such as ANEC are able to implement creative strategies to help mitigate the economic pressures caused by export-oriented agricultural policies. These projects, as seen by the description of Nuestro Maíz, are vulnerable—politically, structurally, and financially. They are also vulnerable to the particularities of local markets as they are replicated across social and economic contexts, even within the same country. Social entrepreneurial projects are perhaps vulnerable because of their extra-economic associations with government agencies and civil society organizations.

ANEC's work to build the capacity of small- and medium-scale farmers in the field as they manage their production practices and organizationally as producer groups, seems essential in order to create resilient food networks. ANEC is "performing the economy otherwise" in that it values the agricultural, cultural, and ecological work of Mexican campesinos. In the organization's attempt to create new markets for maize, it is possible to see biocultural agrifood relations in the making. The shortened food supply chain created through the Nuestro Maíz project both contests and reflects neoliberalism. In this case, social entrepreneurship—an *empresa social*—emerged from the spaces left empty as the state shifted policy priorities away from campesino agriculture. ANEC, in an effort to respond to this shift, is reshaping food networks, reflecting, contesting, and even reinforcing neoliberalism.

Through the Nuestro Maíz case study the biocultural agrifood relations of local food networks are revealed. The Nuestro Maíz maize network is not a static, uni-directional flow of maize between farmers and eaters, but poignantly illustrates the profound interrelationship between nature and culture, how the global permeates the local, and how markets and social movements are intertwined. This circulation and intersection of maize ecologies, economies, and cultures at various scales reflects the "global nature" of maize. The dynamics of diversity are constantly shifting, resulting in competing notions of quality and demonstrating how food networks are constantly made and remade through these interactions. It is among these dynamic and confusing conditions that food networks such as ANEC's Nuestro Maíz are flourishing and failing.

The notion of biocultural agrifood relations offers a lens through which to see how the Nuestro Maíz supply chain is *constructed* as a reflection of the neoliberal context. ANEC's political and productive vision draws on the historical, ecological, and cultural importance of maize to create a shortened maize supply chain linking local farmers to local eaters. Similarly, the biocultural practices of Nuestro Maíz can be *deconstructed*. At the same time as maize is enlisted as an act of resistance—the project uses the cultural politics of maize in Mexico to market the tortillas—this is complicated by consumer demands that have been shaped by global markets. In response, the project enlists the science of hybrid development, in partnership with local farmers participating in the project, reshaping the farmer's agricultural practices. The coming together of two "worlds of diversity" illustrates the ongoing agency of maize as it shapes crop improvement for the project. Understanding this dynamic as *relational* offers a way to view diversity as fluid as opposed to fixed and place-based rather than place-bound. These examples serve to illustrate the challenges faced by social movements engaged in constructing place-based, translocal markets.

Understanding these conditions through a careful examination of Nuestro Maíz and ANEC contributes to my understanding of biological and cultural diversity as dynamic rather than static. Agrodiversity, as illustrated by Nuestro Maíz's value-added tortilla project, shapes and is shaped by technology, institutions, economies, and consumer preferences. Market-based agrodiversity conservation efforts grapple with these dynamic realities, illustrating how static notions of diversity must be left behind. Attention to the ecologies, cultures, and economies of specific places reveals the uneven terrain that must be navigated by food networks in the making, as well as the contradictions and challenges they embody.

Notes

1. For a discussion of "the politics of embeddedness," see Friedmann and McNair (2008) and Gibson-Graham (2006), building on Polanyi (1944).
2. www.laneta.apc.org/anec.
3. Other funders include FIRCO (Fideicomiso de Riesgo Compartido, or Shared Risk Trust) and FIRA (Fideicomiso Instituidos en Relación con la Agricultura, or Instituted Trust Related to Agriculture). All funders are affiliated with SAGARPA (Secretaría de Agricultura, Ganadería, Dessarrollo Rural, Pesca y Alimentación, or Secretary of Agriculture, Livestock, Rural Development, Fish and Food).
4. The close interrelationship between production and consumer units is also discussed in Sally Miller's (2008) work on Maritime cooperatives and illustrates a central characteristic of local food networks.
5. These are based on the prices of maize at the time of my fieldwork (2006–2008), and serve as an example of how value can be added to the maize of the farmers.
6. The larger-scale tortilla plants tend to use maize flour, while the thriving economy of neighborhood-based *tortillerías* continues to use masa.
7. This has to do with the starch and moisture content of the grain, which varies from landrace to landrace.
8. Victor Suarez is the founder of ANEC, and positions the organization as working parallel to the state. Suarez held a seat in the Cámera de Diputados (Mexico's federal congress) from 2000 to 2006.
9. See www.rmalc.org.mx/principales/conv_soberania_alimentaria.htm for more on this national campaign "in defense of food sovereignty and the reactivation of the Mexican countryside."

4

Itanoní Tortillería

> The reinvention of eating a taco in the city made with peasant maize could be one of the many social possibilities that would help reactivate "marginalized" agricultures. If the claim for social development policies were to be reoriented as a means to enable peasant societies to maintain their agrodiversity, their maize could be added to consumption processes of markets beyond local borders.
>
> —Vizcarra Bordi 2006, 104

The Rich Oaxaca Context

The Itanoní Tortillería links urban eaters in Oaxaca, Mexico, with campesino farmers around the state. It is precisely, as Ivonne Vizcarra Bordi notes in the opening quote, this vision of reactivating campesino agriculture as a strategy to conserve agrodiversity that motivates the owners, Amado Ramírez Leyva and Gabriela Fernández Orantes. This chapter begins by describing Oaxaca in order to contextualize the work of Itanoní. The following sections describe the supply chain, the motivations of the owners, and their work with campesinos. The project provides a way to elaborate the notion of biocultural agrifood relations and explore its epistemological roots.

Oaxaca is famous for its ecological and cultural diversity. Topologically the state spans from the Pacific Ocean to the Sierra Madre mountains that reach more than 2,700 meters (over 8,800 feet) in some areas. The mountainous characteristics of the state means that the ecology varies wildly, from dry savannah to tropical cloud forest to coastal beaches. There are seventeen distinct ethnolinguistic indigenous groups that live in the state (Barabas, Bartolome, and Maldonado 2003), speaking over 104 languages and dialects (Toledo 2001). The ethnocultural diversity and topographical variation mean Oaxaca is the Mexican state with the most biodiversity. Oaxaca is one of the regional centers of origin in Mexico of maize, chili peppers, and squash. It also means that Oaxaca has a large number of small-scale campesino farmers who farm over 7 million hectares of land, almost two-thirds of the state's territory (Toledo 2001, 476).

Oaxaca is the second poorest state in Mexico. Agriculture employs 40 percent of Oaxaca's working population. This production is characterized by small-scale family enterprises growing for household consumption. A small amount of surplus production is taken to the market. The land is primarily tilled by bullock, but farmers sometimes pay to have a *tractorista* till their land. Harvesting is done by hand. Oaxaca has a high degree of land tenure fragmentation, with individual parcels ranging from two rows to five hectares (Badstue 2006, 57–75).

Although agriculture remains the key component of household economies in Oaxaca, off-farm work and remittances are also important sources of income. Over 150,000 Oaxacans leave the state to work in Northwest Mexico or the United States each year, and over 1 million Oaxacans live in the United States (CEAMO 2002 in Badstue 2006, 57–75).

Oaxaca City is known as the cultural and culinary center of Mexico. Its colonial architecture, vibrant zocalo, art galleries, artisanal crafts, cafés, and restaurants are attractive to tourists. Over the past ten years a number of cooking schools have opened in Oaxaca, making the city a popular place to come and learn to cook Mexican specialties such as mole. The food markets in and around the city are world famous for the diversity of products they carry. The availability of visually stunning and diverse market products, tantalizing street food, and haute Mexican cuisine offer the international gourmand endless opportunities to explore interesting ingredients, complex flavors, and long-standing culinary practices.

Strong, autonomous indigenous movements, civil society organizations, and student movements have emerged from this rich ecological, cultural, and social context. For example the Asemblea Popular de los Pueblos de Oaxaca (APPO; the Popular Assembly of the Peoples of Oaxaca) coalesced after the brutal repression of a teachers' strike in 2006, and now represents a powerful political and cultural movement of diverse social groups demanding political reform and autonomy (Esteva 2007). Indigenous movements and organizations are strong and active in Oaxaca. Two examples close to this research include the work of CEDICAM (Centre for the Integral

Development of Mixteca Campesinos, or Center for Integral Small Farmer Development in the Mixteca) on environmental and agricultural issues in the Mixteca Alta and the work of UNSOJO (Zapotec Union of Organizations of the Sierra Juárez of Oaxaca) to defend maize agrodiversity in the face of GM (genetically modified) contamination. These social movements and civil society organizations have been at the center of political conflicts and social innovation in Oaxaca, and often they focus on livelihoods, environment, culture, and health. Itanoní is embedded within this context and draws on the cultural and ecological diversity of Oaxaca, complementing the civil society movements. The Itanoní Tortillería is part of an interesting movement in Oaxaca to recover traditions, history, spirituality, and cultural identity (Aldo González, pers. comm., March 2006) and therefore allows for a deeper understanding of biocultural agrifood relations.

The Itanoní Tortillería Supply Chain

Itanoní Tortillería is a family business, started by Amado Ramírez Leyva and Lea Gabriela Fernández Orantes (a husband and wife team). They opened a tortillería in September of 2001 selling tortillas made from four maize landraces. By January 2002 the demand for tacos made with the tortillas and *guisos* (taco fillings) encouraged Amado and Gabriela to open a restaurant in a middle-class neighborhood with a simple menu and table service. In October of 2005 a second restaurant was opened in the historic center of Oaxaca.[1]

The Itanoní Tortillería supply chain can be conceptualized as a full circle of economic activity. Regional farmers grow maize for the *tortillería*, where it is first transformed into masa and then into traditional Mexican snacks—*antojitos*—prepared on a clay *comal*. The connection, however, does not stop with the consumer who eats *antojitos* from the *comal*. The restaurant owners, by creating maize production cooperatives with the farmers, transform the production-consumption chain into a circle. The following description of the supply circle offers an opportunity to follow native maize varieties from farm field to restaurant table in the form of menu items that are carefully chosen to reflect the specific flavors and qualities of each variety.

Maize for the *tortillería* is grown by four farmers in diverse regions of Oaxaca state. The geographical and ethnocultural differences between the regions, as will be explored here, dictate the characteristics of the maize as well as the agroecological and sociocultural practices of the campesino farmers. Amado developed relationships with these farmers during former work selling native criollo seed in the state of Oaxaca. Working in partnership with the farmers, he purchases their maize to use in the restaurant. In this section

I focus on the overall supply circle, leaving the details of these relationships for the following section.

The maize is brought to the restaurant from the farmers' store houses several times a year. Every evening the maize is soaked in hot water with *cal*. Early each morning the maize is ground into masa.

At Itanoní, the open concept of the restaurant encourages the public to observe the *antojitos* being made. Walking through the processing facility, one can see maize soaked, mixed with *cal*, and ground. In the center of the restaurant women press tortillas by hand. The kitchen, tucked into the back of the restaurant, is a hub of cooking activity. There, the *guisos* (taco fillings) and specialty drinks are prepared—*atole* and *tascalate* made from the maize, as well as Oaxaca's famous chocolate. Surrounded by tables, the clay *comal* is the place where all the dishes available at the *tortillería* are prepared. The *guisos* sit in clay pots, ready to be spooned onto tortillas warming on the *comal*.

The restaurant buys fruit and vegetables from the local markets in Oaxaca, and sources products like eggs, chickens, pork, and beef from farmers who treat their animals humanely. Gabriela described how she is willing to pay twice as much for chicken from a woman who cares for her chickens, feeds them well and has enough space for them. "I believe that inputs of good quality give a good quality product." Amado and Gabriela seek out products that will complement the maize. For example, they use *chile de agua*, a chili pepper that is native to Oaxaca. This pepper is quite delicate and needs a lot of care

PHOTO 4.1
The *comal* at the center of Itanoní Tortillería.

to grow. Gabriela buys the peppers directly from women who grow them in their kitchen gardens. Amado and Gabriela also support local entrepreneurs, selling such Oaxacan specialty food products as *chipolines* (fried grasshoppers), amaranth snacks, chile paste, and *tascalate* mix at Itanoní Tortillería.

As in the restaurant business anywhere, the staff at Itanoní are usually transient. Servers and *tortilleras* (the women who make tortillas), are generally from rural areas outlying Oaxaca, new to the city, searching for a job, and are often on their way to the United States to work as migrant laborers. Patricia Bautista Bautista, who was eighteen years old at the time of my fieldwork, worked at Itanoní for four months before leaving for the United States. In an interview she told me that making tortillas reminded her of her childhood in the Oaxacan countryside. She grew up rising early to help her mother prepare tortillas for the family. Patricia told me she was leaving for the United States the following week. Her two-year-old son was staying behind while she embarked on the journey from Oaxaca to the United States border by truck, where she would then endure the dangerous walk across the border. This journey was going to cost her two thousand U.S. dollars, but Patricia was excited and optimistic about her future. Although she didn't mind making tortillas, Patricia told me, she looked forward to doing something different. The repetitive work was causing her wrists to ache.

Photos 4.2 and 4.3 show Patricia kneading masa and heating tortillas on the *comal.*

PHOTO 4.2
Patricia Bautista Bautista kneading masa.

PHOTO 4.3
Tortillas on the *comal*.

Both the Itanoní supply circle and the restaurant staff are examples of how local food networks are embedded within and reflective of broader social, economic, and political forces. The supply circle is partially local, and indeed Amado and Gabriela are working to source more local products and grow what they need at the restaurant in partnership with their network of farmers. One day while working at the restaurant, I questioned Amado about why Itanoní served Coke. He replied with a laugh that you could not have a *tortillería* in Mexico without offering Coke to drink. The staff also reflect the broader political, cultural, and economic forces that shape Mexico and Oaxaca. Women work cooking and making tortillas while men serve tables, reflecting a historical gendered division of labor in Mexico. Migration, as reflected by Patricia's story, is an economic necessity, but also a cultural rite of passage for young Mexicans seeking new experiences.

At Itanoní, these contradictions are only the beginning of the many ways that the *tortillería* stretches beyond Oaxaca, drawing inspiration from other places, linking with other food movements, and grappling with the reality of operating a restaurant in a vibrant, multiethnic, and global city. Itanoní's activities are grounded in the ecologies of each of the campesino farms producing maize for the restaurant, as well as the culinary and cultural traditions of Mexico. The extremely local aspects of the supply circle contrast with the conventional aspects of the supply chain—transient labor, gendered roles, and products produced by multinational corporations. The work at Itanoní is reflective of local food networks in the making, partial expressions of deeper ecological and cultural goals. The maize and culinary practices are grounded in the particular ecologies, cultures, and economies of Oaxaca. But global economic trends permeate the shortened maize supply circle.

In the following section I discuss the entrepreneurial motivations of Amado and Gaby, describing their backgrounds, Itanoní Tortillería's history, and their dreams for the future. I then return to Itanoní's menu as a way to explore the notion of culinary conservation. This leads to a discussion about identity and diversity, Itanoní's work with farmers, the maize used, and the impact of Itanoní's work on maize agrodiversity. I conclude by commenting on the biocultural agrifood relations of Itanoní Tortillería.

Biocultural Entrepreneurship: Building Food Networks

Both Amado and Gabriela have backgrounds working on rural development projects in Oaxaca and Chiapas. Amado is an agronomist and has worked on numerous maize agrodiversity initiatives with the International Corn and Wheat Improvement Center and the National Institute for Forestry, Agricultural and Livestock Research (INIFAP). Prior to opening Itanoní, Amado operated a business that improved maize landraces and sold seed to farmers in Oaxaca. He opened Itanoní to promote maize agrodiversity in the city of Oaxaca, to address the link between agrodiversity and culinary practice, and to support local farmers through the use of their maize.

Amado grew up in Oaxaca City. Important influences in his life were his Mixtec grandparents, campesinos who farmed in the Mixtec region north of Oaxaca City. Amado described his grandfather as a wise man who spoke little but was very patient with his urban grandson, teaching him how to work in the field. Amado stressed how this experience with his grandfather was very important; the childhood opportunity to visit the farm was like a "return to and reconciliation with the earth" (pers. comm., February 28, 2006). Amado's grandmother, who is still alive, has provided Amado and Gabriela with many recipes that are on the menu at Itanoní Tortillería. She was a wonderful cook, with a deep appreciation for good ingredients, described Amado.

Amado began his career as an academic. He did his master's of agronomy in Germany and then taught at the University of Chapingo for several years. Critical of academics working in the field of maize agrodiversity, Amado states, "The Mexican situation is caught in an academic crisis regarding the disconnection between research and what is happening in reality." Amado questions the way in which academic and research networks approach their work—first by seeking to validate their research objectives and second by only doing research for which there is funding. Amado's critique has resulted in a deep commitment to create ways for people to live from maize agrodiversity. After several frustrating years working in academia, Amado felt he could not do the work he needed to do from within the university. Amado described this journey in the following way.

> I really believe that the countryside, especially the Oaxacan countryside, is a place of poverty, misery, pain, death. Maize lives in this environment of poverty and misery. If you live in an environment of pain, your surroundings—the countryside, the people, the way of life—begin to signify pain. For this reason I understand why my mother never wanted to return to being a *campesina*. For her it signified pain, having to get up at four in the morning as a child to grind the masa. At Itanoní we romanticize grinding the maize. But in the countryside grinding masa, making tortillas, making salsa, going to collect wood in the mountains, carrying the wood is all work that signifies pain. When you live this life you do not have a childhood of playing games, like our children now have.
>
> From this you can understand why campesinos are limited in how much they can work for themselves, in preserving their surroundings, because the land around them represents pain. They don't want to preserve something that is painful. We see this at Itanoní. People who come from the country don't want to make tortillas. Tortillas in their world means hard work, heat from the fire, respiratory illness, crying eyes. Well, not only this. There are many positive things as well. But in this context the diversity of maize is interrelated with poverty. This is why we have to break this cycle and begin to value the environment of biodiversity.
>
> As an example, I'll tell you my story. I had to come to terms with the fact that I was going to be a *taquero*. That I was going to make tortillas, and I, who has a master's degree from Germany, who has traveled all over the world, who is a national researcher . . . now a *taquero*? This was a cultural shock. (pers. comm., February 10, 2006)

In this quote, Amado describes his own motivation to start Itanoní Tortillería as a concrete way to break the cycle of poverty in the Mexican countryside. But imbued in this quote is an articulation of the deeper class and racial issues that are tied to Mexican culinary and agricultural history. For many campesinos, leaving the countryside is a way to escape the drudgery of everyday life. For Amado, reclaiming this way of life—becoming involved in agricultural production and making tortillas—is a way to revalue campesino culture and address the poverty of rural communities in Oaxaca. At the same time, however, Amado grapples with his own internalized racism and classism that reflects culturally shaped perceptions about campesinos and *taqueros*. Amado recognizes that these are the broader cultural stigmas that need to be confronted through his work at Itanoní. Recognizing his privilege, Amado articulates that he is in a unique position to be able to leave behind these cultural stigmas and then choose to face and engage with them.

These issues are echoed by Gabriela. By academic training, Gabriela is a chemical engineer with a specialization in aquatic resources. She grew up in Chiapas in a mestizo family. Earlier in her career she worked in Chiapas on a rural credit project supporting fishers as the fishing stocks disappeared. Then she moved to Mexico City where she worked for the National Indigenous Institute. When the guerilla movement gathered strength in Chiapas in the late 1980s, Gabriela worked for eight months with indigenous and campesino peoples in the costal region. Gabriela and Amado met during this time in Chiapas, where Amado was working with indigenous communities, and got married in the early 1990s during the Zapatista movement. Their experience in Chiapas, working with indigenous communities was formative. Amado stated that in Chiapas he observed extremely organized, cohesive, and productive communities, leading him to question why other parts of Mexico did not have the same social, political, and cultural resources.

Getting involved in a *tortillería* business was not an easy decision for Gabriela.

> At the beginning, when we were thinking about opening a *tortillería* as a cultural development project, I didn't want to. I knew what it meant to have a *tortillería*, what it stood for in Mexico, so I didn't want to start the project. I didn't want to be a *tortillera*. Nor did I want to have a restaurant. I knew how much work it implied. My grandmother had a restaurant for 35 years, and my mother had a taqueria for twenty-five years. They worked too hard.
>
> Amado and I talked about the project. I liked the idea, but I didn't want to do it because of the work. Finally, I said all right, let's do it. But, I said, we are going to do it with criollo maize and the traditional process. Clay *comals*. We are going to create the right context and surroundings. And so we started making tortillas. Then we started making beans and salsa so that people, on their way home from work, if they didn't have time, could make a taco and have a quick dinner. We made tortillas from criollo maize, and it was incredible, magical.
>
> We opened at the end of October, and by January and February people were asking, "What should I put on my tortilla?" I told them, "Cheese and beans." They asked, "Where are the tables?" I replied, "This is a *tortillería*, not a taqueria." People kept asking me why it was only a *tortillería* and not a restaurant. I kept saying, "No restaurant." For five days in a row people asked me, "Why don't you have *guisos*?" I kept saying no. I didn't want to have a restaurant. After five days of this I agreed. The clients were asking, so I decided that we had to orient the business this way. I couldn't deny it. I was the only one who was denying it and rejecting the idea of a restaurant. Finally, we started with simple *guisos*. (Fernández Orantes, pers. comm, February 16, 2006)

For both Amado and Gabriela, there were issues related to the type of work involved in starting a *tortillería* and restaurant. For Amado, the issues related to class. For Gabriela the issues related to the gendered work of making tortillas. She had seen her grandmother and mother work in restaurants, and knew the work this implied for herself as a woman. Making tortillas is women's work and is repetitive and strenuous. Although women no longer have to grind masa by hand on a metate, kneading, rolling, and pressing tortillas is still onerous. Gabriela articulated these issues in relationship to her role at Itanoní. She is in charge of the daily purchasing and cooking at the restaurant, while Amado works with the farmers. Gabriela notes that she and Amado talk about this frequently and attempt to share the more interesting work related to their project evenly, as a way to counter the gendered aspects, monotony, and hard work of running a restaurant.

Itanoní Tortillería is described by Amado and Gabriela as a "social entrepreneurial" project. After having participated in many other development projects, they decided that a small-scale project rooted in the market was where they could have the most impact. Amado described the market as holding hope for maize conservation and social change. He sees his work as simultaneously supporting farmers to create sustainable livelihoods options and supporting eaters to understand maize agrodiversity.

Amado and Gabriela view their work to promote maize agrodiversity as distinct from, but complementary to, other activities to promote and conserve maize agrodiversity in Mexico. Gabriela describes the work at Itanoní as parallel to the work of activists who are protesting GM corn in Mexico.[2]

> God put me in a place where I have to *do* something. Our work is separate from the work of the activists. Either you dedicate yourself to protesting *against* things or you dedicate yourself to *creating* something. Our work at Itanoní is a dizzying kind of work that is very inspiring for people. But it doesn't leave much time. I either work here *or* I go to protest. I cannot do both. I am doing what I have been called to do. For this reason, I say that God put me in this place; he didn't put me here to protest. Or it is a different kind of protest, isn't it? (Fernández Orantes, pers. comm., February 16, 2006)

When asked if the two parallel lines of action or activism cross, Gabriela continued,

> I don't think they cross, and I don't think they are going to cross. Because I can only do one thing or the other. To create an alternative, I need to be at the restaurant, and this requires 100 percent of my time. This is a family project; we live this. It requires all of our attention. There is no time to go to picket, or go to a protest march, or go to a conference. If the *tortillera* doesn't come to work, I have to make the tortillas.

Gabriela describes the way the protesters support the work of Itanoní as twofold. First, they support Itanoní by frequenting the restaurant and buying tortillas. Second, they spread the word to people by telling others that Itanoní is an alternative. Amado and Gaby see themselves as part of a broader movement in Mexico, as allies with other social and environmental justice groups. Amado described their work as being aligned with the One No, Many Yeses strategy articulated by the Zapatista movement.

Amado and Gabriela, during the months I worked at their *tortillería*, supported the work of practitioners and activists in several ways. Two occasions stand out. The first was an event as part of National Agronomy Day. Agronomists from around the state of Oaxaca, part of a statewide organization, came to Itanoní to eat lunch. The lunch that was served was a delicious example of the regional agrodiversity of Oaxaca. Chicken in a red mole sauce was served with tortillas of four different colors. Three different kinds of salsa, *aguas* (fruit juice) from native fruit, and *mezcal* (a drink similar to tequila) made from specific varieties of agave were also served. Sitting at a table of agronomists, I heard them exclaim over the food, discuss the tortillas in great detail, and reminisce about the tortillas their grandmothers made. They all saw great value in the work being done at Itanoní. Beyond the wonderful taste of the food, they spoke about the importance of using criollo maize and exposing urban eaters to diverse flavors. One man spoke about Amado's vision to support farmers through his network of restaurants. This, he felt, was critical for developing an economically viable agricultural sector in Mexico.

The second occasion that I observed Itanoní supporting the work of others involved in agrodiversity conservation was when Zapatista leader Subcomandante Marcos came through Oaxaca with a caravan of activists and supporters as part of the Otra Compaña. The campaign was organized during the lead-up to the federal elections that were held in July 2006. Zapatistas traveled across Mexico to generate dialogue about alternatives to the current political system outside the formal electoral process and to galvanize interest and support for these alternatives. Itanoní supplied tortillas to the caravan at their meeting with the nonprofit sector in Oaxaca. Amado wrote a statement of solidarity that Gabriela read to Subcomandante Marcos and the caravan members. Photo 4.4 shows posters announcing the campaign.

Oaxaca is the center of struggles for maize agrodiversity conservation in Mexico, and Amado and Gabriela are integrated into conservation, environmental, and indigenous networks working to preserve agrodiversity. The activities of Itanoní synergistically interrelate with other local, national, and international initiatives. For example, Amado and Gabriela have participated in the Slow Food movement's Terra Madre conference that brings together artisanal food farmers and traveled to Italy in 2004 to meet with other food

PHOTO 4.4
A poster advertising "la Otra Campaña."

communities from around the world.[3] Amado and Gabriela participate in a growing movement in Mexico to link farmers with eaters. They have experimented with selling tostadas to an organic distributor, Arco Iris. This effort, however, did not work out financially. Amado explained that the organic certification process was too expensive and that it was difficult to find packaging for the fragile tostadas.

Other efforts are underway to build local food economies in Oaxaca and Mexico. While in Oaxaca I attended an agroindustrial fair that reflected the agricultural practices of the region. Small community food enterprises were featured, selling value-added products made from locally grown honey, amaranth, agave, nopal, and chiles. Organic farmers' markets are proliferating in Mexico and are distinct from the produce and food markets that have existed in Mexico for centuries. The new farmers' markets provide niche markets for local, primarily organic farmers, linking farmers with eaters, and are networked through an organization based at the University of Chapingo.[4] The organic markets often draw tourists and are quite expensive, but serve as an alternative outlet for organic farmers.

The networks forged by Itanoní are not as clear as the networks created by ANEC (Asociación Nacional de Empresas Comercializadoras de Productores del Campo, A.C.) and other maize conservation organizations. They represent informal, personal networks and because of the pressures of the work at Itanoní, are often neglected. Amado and Gabriela identify this weakness and have plans to develop a foundation to address the more educational aspects of their work. The Foundation for Identity and Diversity is being developed by Amado and Gabriela and has a mission to "promote cultural identity through agrodiversity preservation of maize and other native crops, recognizing the intimate relationship between natural environments, food crops, kitchens, and the table." This mission will be achieved by a series of educational activities related to maize agrodiversity and Mexican agricultural and culinary traditions. The project is in its infancy, but involves a small group of people who are guiding the development of a constitution and working to register it as a nonprofit organization. The scope of work of the foundation will include developing rural educational centers, cooking tours and demonstrations, linking researchers to practical projects related to agrodiversity conservation, and criollo maize crop improvement. The project to blend marketing and education can be seen as a characteristic of the Slow Food movement. Fonte (2006) describes how Slow Food in Italy and Coop Italia are working together to bring to market the products from Slow Food's Ark of Taste Presidia across Italy. To market these products, their quality (and *qualities*) are described to eaters through in-store marketing materials.

George Stevenson differentiates the practical projects of entrepreneurs and farmers working on local food economies from the oppositional work of activist organizations. He describes the former as "builders" and the later as "warriors," both essential to shifting toward agroecological food systems (Stevenson 1998). Itanoní's work is part of a growing movement of artisanal food communities who are building new markets for their products. The notion of food community is described by Carlo Petrini, founder of the Slow Food movement, as "a physically identifiable entity, which holds values, interests and common purposes and which is engaged in seed saving, harvesting, breeding, animal husbandry, farming, fishing, food processing, food distribution, marketing, education and other eco-gastronomic activities, in order to bring small-scale, quality food products to eaters" (Petrini 2005). Itanoní meets the criteria set out by Slow Food for products that are (1) linked historically, socially, economically, or culturally to particular places; (2) meet taste, environmental, and social quality criteria; and (3) are produced on a small scale by small-scale farmers or processing companies (Slow Food 2008). This growing network of artisanal food farmers, processors, and chefs has become a political movement in opposition to global, corporate, fast food, and industrial agricultural practices. The network can be seen as valuing historic biocultural culinary traditions and also as reclaiming these traditions in places where they were lost or never existed.

Itanoní's potential lies in its ability to link agrodiversity conservation with the growing urban interest in local food. Galvanizing urban interest in agrodiversity by focusing on Mexico's culinary culture and the unique flavors of maize landraces is a creative strategy to create an alternative maize economy—one that supports local farmers and agrodiversity—in Mexico.

The personal motivations and networks of Amado and Gabriela shed light on the work of biocultural entrepreneurs building local food networks and linking to oppositional social movements. This is an expression of the place-based but not place-bound work of biocultural entrepreneurs as they forge local food networks, yet connect to broader movements through networks and coalitions. Amado and Gabriela embody the characteristics of "rooted cosmopolitans" described by Tarrow (2005, 29) and Edelman (2008, 232) as activists who are engaged in place-based struggles yet are connected translocally to other social movement activities.

For Amado and Gabriela the challenges of hard work are balanced by a sense of reclaiming ecological and cultural practices and contributing to the regeneration of agricultural landscapes and culinary traditions in Mexico. Itanoní is an experiment of living agrodiversity. Just as "this rice tastes like a lake" for the Anishinaabeg wild rice harvesters around the Great Lakes in Canada and the United States, the tortillas at Itanoní taste like the land it was grown on and the hand that planted it. For Amado and Gabriela the flavor of maize links urban customers and campesinos in the country, a cyclical relationship connecting the past, present, and future through agricultural and culinary tradition.

Culinary Conservation: Eating, Sharing, Learning

Through the work of the Slow Food movement, the notion of culinary conservation has been developed to describe the work of food communities rooted in place, ascribing importance to culinary traditions and techniques. Amado and Gabriela see their own work at Itanoní as focused on *using* maize agrodiversity as a strategy for conservation. In this sense conservation goes beyond thinking of agrodiversity in terms of genetic resources to seeing it as intricately linked to particular ecologies and cultures. This section focuses on Itanoní's menu and the experience of eating at Itanoní. In the following section I focus on Itanoní's work with the farmers who grow the maize. Both sections extend the ideas about agrodiversity as a fluid biocultural process and food networks as place-based but not place-bound.

The dishes served at Itanoní are chosen to highlight the flavor of the different maize varieties used. Regional dishes from around the country are chosen to complement the Oaxacan maize. These dishes are known as *antojitos*, or snacks, in Mexico, and are usually eaten after breakfast between nine and eleven in the morning, in the late afternoon between four and six, or in the late evening. It is common to see a taqueria open at one in the morning and bustling with business. Taco stands are a part of Mexican street life but are being threatened by the incursion of foreign fast food restaurants like McDonald's and Burger King. Itanoní is open from eight in the morning to four in the afternoon catering to the morning crowd and people interested in a quick lunch lighter than the usual *comida*, or midday meal. Many people come to buy tortillas to take home for their lunch.

Tacos and *memellas* are traditional Mexican snacks made from maize. A taco is a tortilla with a topping, or *guiso*. A *memella* is a thicker, smaller version of a tortilla typically topped with pork lard, beans, and cheese. The *tetella*—a tortilla stuffed with beans and cream then folded into a triangle—is a snack from the Mixtec region of Oaxaca and a recipe shared by Amado's grandmother. The *ese* is a Chapaneco snack from Chiapas that Gabriela remembers from her childhood—a tortilla with an *oja de santa*[5] pressed into it, and then filled with beans and cheese. Also available at the restaurant is pozole, a soup made from large maize kernels, and black bean tamales made by one of the campesino families associated with Itanoní. Gabriela and Itanoní's chef, Sabina, are always experimenting with new recipes to complement the maize. They seek to offer common Mexican snacks, but also to recover, reclaim, and promote Mexican culinary specialties.

Itanoní has created a small niche market for maize landraces, geared toward urban eaters. The people who eat at the restaurant are generally middle-class urban Oaxacans. Many tourists are beginning to learn about the restaurant, so there is an increasingly international clientele. Amado describes his clientele as 60–70 percent local residents who come because of the flavor of the food and 30–40 percent activists, students, chefs, and culinary tourists looking to experience an interesting project. Most eaters understand that they are eating maize landraces grown by regional farmers. They learn this through educational materials that are visible at the restaurant—a banner that hangs at the front of the restaurant and placemats that tell a story of the cultural importance of maize in Mexico.

Itanoní is catering to, and attempting to nurture, an urban market that appreciates Mexican culinary specialties, in particular, maize diversity. Amado and Gabriela are engaging in a cultural politics of agrobiodiversity based on use-value as opposed to exchange value. Amado distinguishes this as the difference between a focus on the cost of production (use-value) and market price determined externally (exchange-value). The work of Amado and Gabriela promotes an ethics and aesthetics of diversity, while building a local food network that connects farmers and eaters.

Central to the work of Itanoní, according to Amado, is the interrelationship between identity and diversity. These abstract terms are described in the following way by Amado. Drawing on Marxist notions of "essential relations," Amado defines identity as the sociocultural, ecological, and genetic relations that make up a being. Diversity, then, is the coming together of two identities, in infinite possibility. When two maize varieties cross, diversity is perpetuated. In the milpa, describes Amado, maize, beans, and squash interact, sustaining diversity. At Itanoní the eater comes together with the maize resulting in a growing consciousness about diversity. The menu attempts to combine flavors that best express this diversity.

A banner at the restaurant's entrance reflects Itanoní's goals to cultivate an appreciation for maize agrodiversity and the culinary traditions of the region, as well as support local farmers (see photo 4.5).

The banner reads,

> Itanoní . . . maize flower, the art of cooking maize.
> Native maize selected from Oaxaca.
> By consuming it you will have the opportunity to:
>
> Digest the pure flavors of maize developed in our diverse soils and villages.
> Contribute to the preservation and reproduction of diversity: the biological, cultural, and culinary diversity of Oaxacan maize.
> Ensure fair prices for the campesinos who cultivate our maize.
> Taste maize products made with artisanal methods and processes.

The banner is not the only way that Amado and Gabriela share the story of maize agrodiversity in Mexico with their customers. There are colorful posters and photos on the wall that address various aspects of agricultural biodiversity in Mexico. The restaurant has developed some marketing materials, including cards that showcase the varieties used and a placemat that tells a Mexican myth about the creation of maize.

Amado describes the purpose of these educational materials as efforts to raise urban consciousness about maize agrodiversity in Mexico.

> Like globalization, problems are global and solutions need to be global. There are no partial solutions. Preserving the biodiversity of the country has to do with developing consciousness. This is the essential point. How do we achieve consciousness? Well, my response is by eating. If you are capable of being a person who eats consciously, your body will learn to use your consciousness. Profound solutions have to do with what is profound about humanity—the spirit, your unconscious. (pers. comm., February 10, 2006)

PHOTO 4.5
The banner in front of Itanoní Tortillería.

In this quote, Amado connects the bodily act of eating to shifting consciousness, challenging the mind/body dualism, as well as suggesting that conscious eating—consuming biodiversity—will result in a transformation of humanity and agrodiversity preservation. The spirit of maize, *something* beyond its nutritional and cultural importance for Mexicans, will infuse the body and, along with the educational materials provided at Itanoní Tortillería and the traditional processing and cooking methods used, will contribute to a deeper understanding of maize agrodiversity and a more intimate connection between farmers and eaters. This follows Wendell Berry's statement that "eating is an agricultural act" (Berry 1990, 126), connecting eaters to the land and to the agri*culture* of that which is consumed. This intimate connection between the land and culinary traditions contributes to my notion of biocultural agrifood relations and points to the epistemological and ontological shifts required to sustain diversity. The bodily act of eating helps to perpetuate a fluid notion of diversity, one that is grounded in the land but also reflects ever-changing culinary traditions.

Central for Amado and Gabriela is the transformational *taste* of maize. Nurturing a nuanced palate that can distinguish different flavors, or reclaiming embodied knowing, is a central goal for Itanoní Tortillería. Given the percentage of the global population living in urban areas, in particular in Latin America, Amado and Gabriela feel that it is increasingly important to connect urban residents to food and agriculture issues. Urban consciousness about agricultural biodiversity is essential to the creation of a viable Mexican maize economy. Amado and Gabriela work to educate eaters about different flavors of maize, connecting these flavors to distinct criollo varieties, the specific ecologies that the farmers are growing in, and Mexico's diverse culinary traditions. Itanoní's taste education counters the homogenized products of monocultural global production that are responsible for the tremendous loss of agrodiversity.

Despite these articulated goals, the educational aspect of Itanoní is the least developed part of the business. Amado and Gabriela understand the importance of sharing information with their eaters, but have not had time to further develop an educational program that accompanies the eating experience. As a start, a monthly tasting party is held, featuring different maize to taste, accompanied by *guisos* that complement the flavor, as well as other regional products, such as *mezcal* made from particular varieties of agave, juices made from wild fruit. Other ideas described by Amado and Gabriela for developing the educational aspect of their work include mounting a photo exhibit, featuring more native and wild foods on the menu, creating games for children to play that expose them to maize history and culture, and offering farm tours.

Gabriela states that the biggest challenge for Itanoní, one that will have the greatest impact on maize agrodiversity in Mexico, is to raise awareness among children.

> The education needs to begin when we are children. In the case of Mexico, this is a deep cultural knowledge that needs to be recognized and nurtured. I believe that genetically we have an understanding of maize. Children have to grow up with this, be immersed in this, be deeply involved in this environment and culture of maize. At the *tortillería* children have asked me how I dye the maize. I have to explain that I didn't dye it, that maize is like people. There are whites, yellow people, blacks. This is how I explain it. This question is understandable for a foreigner, but from a Mexican? It is terrible that some children don't know the varieties of maize. It is an incredible challenge that we have. (pers. comm., February 16, 2006)

The customers I observed at Itanoní had diverse reactions to the restaurant's food. Occasionally, customers did not like the food. I overheard comments that the tortillas were too tough and the *atole* was not smooth enough (not strained of maize fiber). Sometimes customers laughed at these characteristics of the dishes in a way that evoked a classist bias toward "country" food. Most of the time, however, customers reveled in the hearty flavors of the food, speaking nostalgically about the taste of their mother's or grandmother's tortillas. Many were regulars at the restaurant. Even to someone new to the taste of Mexican maize, the distinct difference in flavor between store-bought tortillas and the tortillas made at Itanoní is striking. It was my observation that customers had a sense of the culinary goals and objectives of Itanoní, but did not know the extent of the restaurant's work with local farmers.

A sense of nostalgia is common among those with whom I spoke about maize conservation. In fact, in many food projects there is a sense of the traditional being better than the modern. The tension between the traditional and the modern is nuanced. Friedmann has explored this tension, illustrating the historical, contingent, and adaptive characteristics of what is often considered "traditional" (2003). This sense of history, and of historical change and adaptation, is captured by the notion of retro-innovation used by Kevin Morgan, Terry Marsden, and Jonathan Murdoch (2006). Here, the focus on innovation is centered, with "retro" implying a reaching back and bringing forward. The work of Itanoní is extremely innovative as a social and ecological enterprise, yet draws on traditional tortilla-making technologies, techniques, and recipes, adapting them to the current social, cultural, and economic context of Oaxaca.

Victor Toledo, when describing the importance of indigenous agricultural practices for agrodiversity, uses the term *historical synthesis* to refer to the "combination of elements of pre-Hispanic origin (such as maize fields, forest products and fishing) with crops derived from European contact (coffee, sugar cane, citrics, or cattle raising) and even 'modern' practices and inputs (such as the use of improved seeds or agrochemicals)" (2001, 478). The notion of historical synthesis acknowledges agrodiversity and culinary practices as reflective of historical processes—fluid rather than static.

Itanoní's work is part of a growing movement of market-based efforts to conserve agrodiversity. Beyond the Slow Food movement, bringing regional and specialty food products to market has been a strategy for biodiversity conservation, environmental protection, and rural development for some time. For example, in *Developing Markets for Agrobiodiversity: Securing Livelihoods in Dryland Areas*, Alessandra Giuliani describes efforts in Syria to bring to market neglected and underutilized species in order to improve rural livelihoods, diets, and ways of life (Giuliani 2007). In Oaxaca, Mexico, a database of non-timber forest products is being developed as a way to document biodiversity and identify economic opportunities for rural communities beyond the forestry sector (Ganz 2003).

But what is the potential for agrodiversity conservation through the market? When posed this question, Amado became philosophical about maize agrodiversity. His perspective on this issue is fascinating and connects his passion for taste and diversity. Amado describes maize agrodiversity as an infinite flow of diversity reflective of the interrelationship between individual maize plants, farmers, and the land. The concepts of flow and infinity challenge static notions of diversity and even challenge the possibility of categorizing criollo maize. According to Amado, maize diversity is constantly changing and adapting to the everyday realities of farmers, their relationship with the milpa, farming practices, agricultural and rural transformation, and changing ecologies.

This notion of infinite diversity clashes with the notion that agrodiversity can be preserved ex situ—in seed banks. Amado's ideas also challenge in situ conservation strategies that are commonly recognized as the best alternative to seed banking. When I asked Amado about the work of his well-known colleagues at INIFAP in Oaxaca working on maize agrodiversity conservation he replied, "Their work is very important. They are classifying the criollo maize. This documentation is part of the necessary work. But it does not recognize the interrelationship between the farmers and the maize. It does not allow for the specific qualities of these relationships to be revealed" (pers. comm., February 10, 2006). Amado's focus on the qualities of the relationships between maize cultures, economies, and markets speaks to a deep biocultural connectivity between the field and table, and "all our relations," human and nonhuman in between.

These multiple, shifting, and fluid notions of diversity point to the complexity of agrodiversity conservation and the multiple approaches required to address the tremendous loss of agrodiversity the world has experienced since the 1950s. Itanoní's work focuses on recovery and restoration of the connections between maize, the land, farmers, and eaters. In order to further understand the relationships between the cultures, ecologies, and markets that shape maize agrodiversity, I turn now to an explanation of the farmer network created by Amado and Gabriela to supply maize to Itanoní.

Identity and Diversity through a Campesino–Itanoní Tortillería Cooperative

Amado and Gabriela have a network of four farmers growing eight varieties of maize for the restaurant. The relationships with these farmers, located around the state of Oaxaca (see map 4.1), was developed through Amado's previous work growing and selling maize seed. The farmers were chosen because they were growing criollo maize varieties Amado liked—good quality, intensely flavorful, and suitable for the dishes offered at Itanoní Tortillería. Each campesino family owns three to five hectares of land. Some land is also rented to produce the maize. One of the campesino families uses irrigation techniques. The rest depend on the seasonal rains.

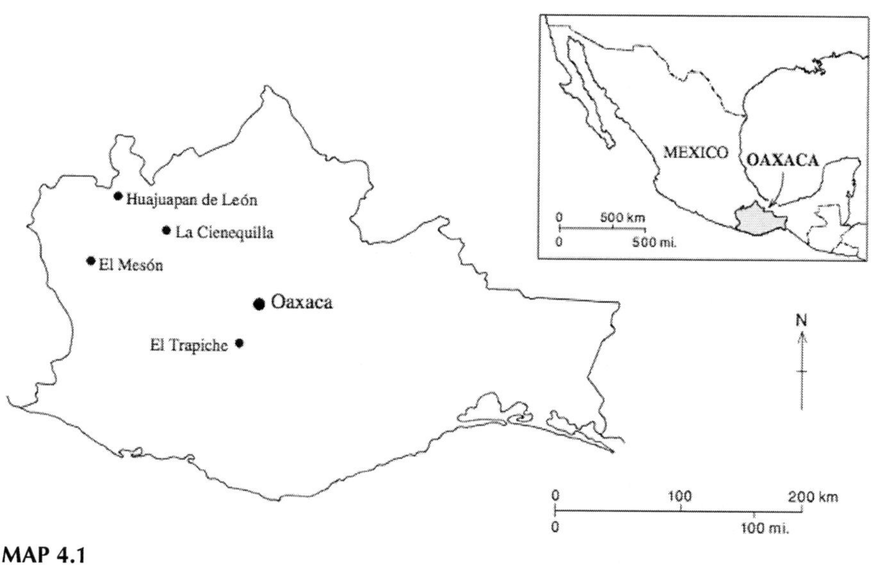

MAP 4.1
Map of Itanoní Tortillería farmers' farms. Map by Carolyn King.

When asked how he initially chose what varieties of maize would be on the restaurant's menu, Amado described his different relationships with the campesinos and the maize varieties they each grow. Pragmatically, the different varieties suit the different dishes made at the restaurant. Some varieties are better for *atole*, some for tortillas, and some for pozole. Amado describes the process of choosing the campesinos and maize similar to the process of "choosing" friends. One does not necessarily choose one's friends; instead, the friendships develop and are initially based on having something in common or a shared event, in this case a maize variety that Amado liked. Amado got to know the farmers that grow for Itanoní through his previous work selling maize seed, developed sound relationships with these farmers, and formed partnerships with them to provide maize for the restaurant.

The network of campesinos is a loose affiliation of campesino-Itanoní cooperatives. The farmers, although part of the informal network, are regionally isolated. They have individual relationships with Amado. Occasionally, there are opportunities for one of the farmers to visit another farmer or meet at a special event. Due to their proximity to Oaxaca City, two of the campesinos have more contact with Amado and Gabriela. I will focus on the relationships between these two campesino families and Itanoní Tortillería.

Amado and Gabriela are currently forming cooperatives with the campesino families who grow maize for their *tortillería*. These cooperatives formalize a relationship that has been established for many years. Amado has purchased maize from these campesinos, worked with them to select and improve their seed, and provided organic inputs. The cooperative structure formalizes and extends this relationship and allows for the campesinos to work with Amado and Gabriela to grow more products for the restaurant. In some respects the cooperative structure works like community shared agriculture, where eaters share the risk of farming with the farmers by paying up front and dividing the harvest later.[6] Once the costs of production, including labor, are covered, the profits from the maize and other products produced through the cooperative and sold to Itanoní are shared equally among cooperative members. The prices for the products are mutually agreed upon each year, based on the market and the costs of production. This allows for campesino families to receive a premium for their maize, and allows the cooperative to avoid the fluctuations of the market. The production methods of the farmers growing for Itanoní are not certified organic (because certification is too expensive), but could be considered ecological as no petro-chemical pesticides or fertilizers are used.

Amado and Gabriela can afford to pay a premium for the maize, due to the direct relationship between Amado and the farmers. This also enables Amado to control the quality and varieties of maize available for the *tortillería*. The campesino families are supported through difficult production cycles. Chucho's harvest in 2005, for example, was destroyed by a persistent drought.

Amado and Gabriela's support (paying for his labor and input costs) encouraged him to continue planting his maize, despite the difficult conditions and a job off the farm that ensures a steady income.

A description of two of the campesino-*tortillería* cooperatives offers more detail about the model used by Amado and Gabriela and is illustrative of the specific ecologies, cultures, and economies that shape maize agrodiversity.

Hector and Luisa—El Trapiche, Zimatlán, Central Valleys of Oaxaca

Hector Díaz Castellano and Luisa Castellano Ortega are the closest farmers to Oaxaca City. Their farm is located in El Trapiche, municipality of Santa

PHOTO 4.6
Hector and Luisa with the maize they grow in El Trapiche, Oaxaca.

Cruz, district of Zimatlán, about forty kilometers from the city. Hector and Luisa consider themselves "people of maize." Their ethnocultural background is mestizo with Zapotec origins. Hector and Luisa are both originally from the Zimatlán district in the Central Valleys of Oaxaca.

The Central Valleys of Oaxaca are three highland valleys that stretch out from Oaxaca City. This region has been the center of Zapotec culture, the largest indigenous group in Oaxaca, for over three thousand years. The climate in the fertile Central Valleys is mild, with average temperatures between eighteen and twenty-two degrees Celsius and five hundred to one thousand millimeters of precipitation annually. The rainy season is from May to October. Much of the area is irrigated. The Central Valleys of Oaxaca are densely populated with 878,132 people (CEAMO 2002 in Badstue 2006, 57–75), and the land is farmed intensively for family consumption and the regional market. Crops grown include maize, squash, some vegetables, and the region's famous tasty black beans.

The proximity to Oaxaca City means that the Central Valleys are exposed to urban life and globalization. Despite this, many aspects of indigenous culture remain intact, including language, food, handicrafts, dress, and community life (Badstue 2006, 57–75). Indigenous social institutions such as *compadrazgo* and *tequio*[7] are still part of family and kinship relations, and reflect values of mutual help, reciprocity, and confidence (Badstue 2006, 57–75).

Hector and Luisa have four children, ages thirty-eight, thirty-seven, thirty-two, and twenty-seven, and four grandchildren. Three of their children live in El Trapiche and one is currently living in the United States. Two sons work on the farm and live with Hector and Luisa. One daughter lives at home. The daughter in the United States lives in California and used to work in a restaurant, but now takes care of her children. Her husband works in construction. They send remittances to Hector and Luisa. Another son used to work in the United States but has returned to El Trapiche to join the cooperative. Hector and Luisa have extended family in El Trapiche, as well as brothers and sisters in the United States.

The three hectares the family farms were previously farmed by Hector's father, who left the land to Hector. The land was part of an *ejido* until 1995 when the *ejido* was dismantled and privatized through the PROCEDE (Programa de Certificación de Derechos Ejidales y Titulación de Solares) program. For Hector this means more control over his production, but less community support during the busy production seasons. The land is irrigated, making two or more plantings possible each year. The family's houses and garden are located a short walk from the fields they cultivate. Surrounding their house is space for animals—chickens, goats, and pigs—as well as a kitchen and ornamental garden where herbs, greens, and some vegetables are growing. There

PHOTO 4.7
Hector and Luisa's land in El Trapiche, Oaxaca.

are several buildings that provide living space for Hector, Luisa, and three of their children and their families. An outdoor kitchen is shared by everyone. A new house is being constructed, financed by the money earned by their son while in the United States.

The newly formed cooperative with Itanoní includes Hector, Luisa, and their sons Isai and Philipe; Philipe's wife, Laxi; and Amado and Gabriela. Currently, the cooperative produces maize, black beans, tamales, and worm castings, but production is expanding to include herbs, tomatoes, tomatillos, squash, other vegetables, chicken, eggs, and maybe pork. Each member of the cooperative will be in charge of a different aspect of production. For example, Amado, Hector, and Hector's sons will manage field production; Luisa, Laxi, and Gabriela will manage the animals, herbs, worm compost, and tamale production. The cooperative strives to re-create the close social relations formalized by *compadrazgo*, where families and close friends are tied together through loyalty and assistance (Badstue 2006, 57–75).

Currently, Hector and Luisa get 30 percent of their income from sales to Itanoní. Hector works as a tractor driver for other farmers with larger tracts of land when there is a demand for his services. The couple sells beans and vegetables. Luisa makes tamales, *atole*, and other snacks that are sold in the town nearby. They also receive remittances from their daughter in the United States.

While in Oaxaca, I had the opportunity to visit Hector and Luisa's farm two times. Both times, Amado had organized groups of international students working in Oaxaca to visit the farm and help with husking maize. In exchange for the students' labor for the day, Hector, Luisa, and Amado provided a simple, delicious lunch for the workers. Labor is in short supply in the Oaxacan countryside. Many of the men of working age have migrated to the United States, leaving a shortage of workers to help with the harvest, husking, and degraining of the maize. Younger men do not want to farm because they cannot make enough money, notes Hector. Hector is close to sixty years old. His sons help out on the farm, but this is more and more unusual, as most younger men search for off-farm jobs. Hector expressed deep gratitude and seemed amazed that Amado was able to enlist the help of so many young foreigners for one day's work. Although I am sure that the students did not work as fast as locally hired help might, the labor was free and cheerful. The days working at the farm were memorable and represented an opportunity to learn about rural Oaxacan life and participate directly, as one student told me, "in maize production and agricultural biodiversity." Amado is organizing more student visits to the farm to experience rural Oaxacan life and help out with various tasks.

Hector grows four varieties of maize. One of these is a regional criollo variety, typically grown by campesinos in the area. This variety is not sold to Itanoní but used by the family to make tortillas. The other three varieties were given to Hector by Amado to plant. Amado explained that many of the criollo varieties in the Zimatlán area have been replaced by hybrids, due to the region's proximity to Oaxaca City and the fact that the area is irrigated. The fertile valley has been dominated by industrial agricultural practices using hybrid, input-dependent seeds. Amado reintroduced varieties that would have grown in the region and that are suitable to the climate and ecology. These include Belatobe, a rose-colored, drought-resistant maize used for tortillas and *tlayudas*. The Amarillo variety is grown for tortillas and maize custard. Finally, Amado introduced a Negra criollo with a short growing cycle suitable for *atole*. Amado wanted a good variety for *atole* in the restaurant, but also knew that Luisa sold *atole* in the nearby town. Amado collected these seeds from other farmers in Oaxaca when he had his seed saving business. The story of the Negra seed reflects how the seeds adapt according to the particular contexts. Amado gave some of the seed to an agronomist who improved the seed's yield over several growing seasons. When Amado tasted the improved maize, it was completely different. The selection for yield had dramatically changed the flavor of the maize. This is an example, states Amado, of how maize expresses the local ecology and production practices.

Amado's approach to seed improvement is not based on improving yield. Amado assists the farmers with seed selection annually. Seed selection is typically done by farmers after the harvest, from the husked cob. Amado prefers to select seed on the maize stalk, as this indicates not only the quality of the grain, but also the characteristics of the plant growth. When selecting for crop improvement, Amado examines the whole plant in relationship to the other plants in the field, as well as the cob. This way, a broad range of characteristics can be observed, Amado explained.

Hector described how he is learning about crop improvement from Amado.

> Criollo maize has been conserved by our ancestors. With Amado we are working to improve the criollo maize. Each year the criollo mixes with the others around it. Before I started working on this project, I didn't know how to distinguish the different criollo varieties. All I saw was maize. Now, working with Amado, I am beginning to see how we can begin to select the criollo. Once in a while we find maize with a rare characteristic. This catches our attention because it is different from the others. It is expressing an ancestral gene and we try to capture the trait. It is important to keep producing criollo maize. Apart from conserving criollo seeds, we want to conserve customs like cooking, and conserve many of the other agricultural practices disappearing from our communities.
>
> Maize is sacred for us. When we harvest criollo maize, we may not perform a ritual or anything, but we are filled with reverence and embrace the maize. We hold the cob to our heart and care for it because it is such an integral part of our life. It is part of our struggle. (Diaz Castellano, pers. comm., February 15, 2006)

This quote not only explains the process of crop improvement, but also reflects the cultural loss of skills and knowledge related to maize production that has been experienced by Hector, and more widely in the region, as hybrid, high-input maize production has become the common practice. Hector is able to acknowledge this loss, while expressing the cultural importance maize holds for him. With Amado, Hector is reclaiming the skills of intercropping, ecological production, and criollo maize crop improvement. Amado, on the other hand, expressed how much he has learned from Hector about production planning, crop rotation, and the region's Zapotec culinary traditions.

The following description of another campesino cooperative working with Itanoní stands in contrast to the regional characteristics of Zimatlán and Hector's farming experience.

Chucho—Tooxí, Nochixtlan, Mixteca Alta

Chucho Reyes is of Mixteca origin, from the Zona Alta, approximately one hundred kilometers north of Oaxaca City in the mountains. The Mixtecs were one of the major Mesoamerican civilizations, never conquered by the Aztecs and fiercely resisting Spanish colonization. The Spanish deforested the region

PHOTO 4.8
Drying maize and squash near Chucho's farm in the Mixteca Alta.

and introduced goats and sheep, which led to erosion. The region suffers severe erosion, with over 83 percent of the land degraded (Dahl-Bredine 2006). Despite the environmental and cultural impacts of colonization, the Mixteca language, traditions, and agricultural and culinary practices persist. In recent years this has led organizations such as CEDICAM to work on environmental and rural development projects that reclaim indigenous practices to restore the ecosystem, increase and diversify agricultural production, and generate sustainable livelihoods.[8]

Because of the isolation of the community and the mountainous topography of the area, agricultural practices historically used are still common practice. Culturally, the community is quite isolated, and many rural traditions persist. These include *mayodomías*, the person chosen to organize community, religious celebrations, and festivals. *Tequio*, or a formalized communal work system, is also in place in the Mixteca Zona Alta.

Chucho lives with his wife and young son. He works full time, at night, as a security guard for the local *diconsa* (subsidized food store). His wife works in the home. They farm one hectare of land. The growing season is short and dry, due to the high altitude.

The newly formed cooperative with Itanoní includes Chucho; his wife, Imelda; his wife's uncle; two of Chucho's brothers; and Amado. Together they will produce maize, squash, chickens, eggs, and worm compost. Currently, Chucho receives a very small amount for the maize he produces, but

PHOTO 4.9
Chucho's land in the Mixteca Alta, Oaxaca.

is interested in expanding his production for Itanoní so that agriculture can form the majority of his income.

Chucho grows a Chalqueño criollo, a white-grained variety that was grown by his father and grandfather and is common in the region. At Itanoní this maize is used for tortillas. Chucho plants his maize as basin maize, or *maíz de cajete*, a technique that conserves water. Shallow troughs are drawn in the soil, then a long stick is used to poke a hole in the ground where a maize seed is placed and covered. The troughs collect rain, and the technique preserves the soil structure and moisture (Reh 1939).

Amado describes Chucho's production practices as deeply reflective of historical cultural traditions in the community, the ecology of the area, and the community's distance from the market.

Reflecting on the agricultural practices of Hector and Chucho, various factors affecting maize agrodiversity can be articulated. Maize diversity is embedded within and inseparable from the local ecology, agricultural practices, sociocultural and culinary practices, and the market forces shaping a particular region and campesino family. These sociocultural and agroecological processes, combined with gene flow from different fields, communities, and regions, determine the identity and diversity of maize. The complex interactions with farmers, processors, and eaters along the food network mean that maize agrodiversity is shaped with each interaction, just as this agrodiversity, in turn, shapes the food network.

The Biocultural Agrifood Relations of Itanoní Tortillería

David Lind and Elizabeth Barham, in their article "The Social Life of the Tortilla," use the tortilla to illustrate historical and ongoing material and symbolic discursive shifts related to maize in Mexico. They suggest that the commodification of food is contestable and contested by everyday practices (Lind and Barham 2004, 58). Itanoní is one example of how commodity markets are not homogeneous, and how people are claiming space for alternative economic activity (Leyshon, Lee, and Williams 2003) that embodies social and environmental values. At the same time, Itanoní's activities are embedded in global capitalist processes that shape, for example, migration patterns and the transformation of the Mexican agricultural sector.

Itanoní's activities characterize biocultural agrifood relations, illustrating how the practices and politics of local food networks can reflect an epistemological shift: to cyclical and relational thinking that takes both humans and nonhumans into consideration; to economies based on social and ecological reproduction; and to a cosmovision that links cultural identity to agrodiversity and collective struggles.

Notes

1. The restaurant in the historic center that opened in September of 2005 is now closed. The location had ongoing problems with the landlord. Promotion of the restaurant was difficult due to a restriction on signage on historic buildings that was caught up in municipal bureaucracy. The restaurant closed in the summer of 2006 because of an ongoing occupation of the zocalo by striking teachers only blocks away. The strike escalated into a serious conflict between the state government and a diverse coalition of protesters in June of 2006. In October 2006 the federal government was called into quell the protest, resulting in a further escalation of violence and repression that received international attention. The Itanoní restaurant in the historic center of Oaxaca, like many businesses in this charming, touristy area of the city, was forced to close. Amado and Gabriela are supportive of APPO, but not active in any of the organizations participating in the movement.

2. In the quote that follows, Gabriela refers to the many protests in Oaxaca related to maize and food. For example, the GM corn issue resulted in several large protests in Oaxaca City, the Commission for Environmental Cooperation presented its report in Oaxaca City, and people protested the proposed opening of a McDonald's restaurant in the zocalo. But Gabriela is also referring to protest as a strategy beyond particular protests.

3. See www.slowfood.com.

4. The Mexican Network of Tianguis and Organic Markets.

5. The common English name for *oja de santa* is "sacred herb." The Latin name is *Piper auritum*. It has an aroma similar to anise.

6. This similarity was identified by Amado during a trip to Toronto in May 2008, where he had the opportunity to visit a number of farms in Ontario, including several using the community shared agriculture (CSA) model.

7. *Compadrazgo* refers to kinship relations that strengthen and build webs of obligation. Translated as godparenthood, *compadrazgo* implies both social and economic responsibilities between families and individuals. *Tequio* is unpaid, obligatory community service common in rural communities in Mexico (PANOS Institute 2002).

8. Jesús León Santos, director of CEDICAM, won the Goldman Environmental Prize in 2008 for his work in the Mixteca. See www.goldmanprize.org/2008/northamerica.

5

The Michoacán Centre for Agribusiness

Markets, Movements, and the State in Michoacán

THE WORK OF the Michoacán Centre for Agribusiness (MCA) is expansive, encompassing and coordinating many different initiatives related to local agricultural economic and community development. Over one day's work, Fulvio Gioanetto, the program manager, may stop by a high school that has built a small greenhouse as part of its extracurricular horticultural training and social enterprise project, visit officials at a local municipality interested in developing an agritourism initiative, or give technical advice to a farmer growing organic blackberries.

In 2002 an alliance between farmers and the government was forged to support social enterprises related to agriculture. This alliance was the predecessor of the MCA, formed in 2004 to provide support for local farmers and develop their market linkages. The MCA has an arm's length relationship with the state government, which allows it a degree of autonomy to develop programs. Despite a relatively small budget (5 percent of the total agricultural budget for the state), staff collaborate with other government programs and departments to leverage a broader impact.

In 2008, Jose Trinidad Diaz Vilchis became the director of the MCA, beginning four years of work supporting farmers growing for the local or regional market with an emphasis on organic and agroecological management systems. MCA's work focuses on community economic development. Jose understands MCA's work as part of a broader social transformation, which must be supported by technical and market transformations. The community

BOX 5.1

In 2008, after both doing research on global-local food networks in Mexico, my colleague Amber McNair and I invited two Mexicans working on food system reform to Ontario for a tour. We, after all, had benefited greatly from our trips to Mexico, developing a deeper understanding of a new context and absorbing many interesting ideas stemming from the work on food issues happening across the country. Amado Ramírez Leyva, owner of Itanoní Tortillería (see chapter 4), and Fulvio Gioanetto, who worked for the Michoacán Centre for Agribusiness, came to Ontario for ten days. We toured the many urban initiatives in Toronto: FoodShare; the Stop Community Food Centre; urban agriculture and community garden sites; ChocoSol, a horizontal trade enterprise; and Local Food Plus, a local sustainable certification agency. Outside of the city we went to Everdale Environmental Learning Centre, Whole Circle Biodynamic Farm, YU Ranch and the Norfolk Alternative Land Use Pilot Project, Prince Edward County, and Niagara—all places in southern Ontario where the local food system is dynamic and innovating.

One year later, I was invited to Michoacán to participate in an Organic Agriculture Diploma course organized by the Michoacán Centre for Agribusiness. Alvaro Venturelli, owner of Plan B Organic Farm in Ontario, and I went to Mexico for a week to share our experience building the local food movement in Ontario and Canada, and to learn about similar work being done in Mexico. These two visits were the start of an exchange between Ontario and Michoacán.

In June 2010, Jose Trinidad Diaz Vilchis and Fulvio Gioanetto visited Ontario and participated in meetings with groups and individuals working on local food system development. Three proposals emerged from these meetings about how to pursue a meaningful collaboration between Ontario and Michoacán: develop short supply chains to import tropical fruit directly from Michoacán farmers; develop a Michoacán-Ontario agroecological network to share information and experience related to developing small enterprises that create local inputs from local plants and micro-organisms; and continue the local food systems exchange.

Chapter 5 is drawn from my experience and observations over two trips to Mexico between 2009 and 2010.

economic development strategy is two pronged: supporting local agricultural and economic development, and promoting access to international export market opportunities. Some examples of the kinds of initiatives the MCA is involved with include developing local, statewide distribution centers for the products of local organic farmers; supporting the transition to organic agriculture; promoting local organic products to local, national, and global markets; developing a farm to school distribution network; criollo maize research and promotion; technical agroecological education of local farmers; supporting a council of organic farmers and stakeholders; and developing a beekeeping network.

This book examines local food markets in Mexico by delving into three local food and maize networks. A number of themes emerge from the initiatives, which are carried through in this chapter. Issues of scale relate to an ongoing conundrum in the food studies literature. Can local food initiatives be scaled up to have more impact and scope, to include more people and provide more food? By looking at the role of the state, some of these questions are explored. Other themes include the possibilities for (and limitations of) expanded notions of the economy and how to create organizational and governance structures that embed food and farming in appropriate ecological, cultural, social, and economic systems. The work of MCA, reviewed in this chapter, offers some insight into these issues.

In previous chapters, the Itanoní Tortillería provided an example of an independent, entrepreneurial small business that has developed a network of farmers growing criollo maize for the restaurant. Nuestro Maíz is a social enterprise launched by a producer organization interested in exploring new market opportunities for local campesino farmers. In this chapter, the MCA is used as an example of how state agencies are engaging in the development and support of local food enterprises and networks. Through a description of their activities, I demonstrate how the state has an important role to play in the diversification of markets, creation of new market linkages, promotion of institutional procurement, and development of the infrastructure required to scale up local food systems. Governments can also play a key role supporting the transition to organic and ecological production practices, in particular in the research and development required for agroecological systems. Many of the activities the MCA supports illustrate the key role the state can play in making linkages across social, economic, and agricultural issues to forward a more holistic or integrated food, health, and agriculture policy.

The unique Michoacán context that gave rise to the MCA is briefly described in the first part of the chapter, as is the history of the organic agricultural sector in Michoacán. The social and economic factors that led the MCA to focus on local food networks are outlined. The following section delves more deeply into "the local turn" in Michoacán, explaining why the MCA began supporting the development of local markets for local farmers, shifting away from their long-standing focus on export markets. The next sections describe the core initiatives and introduce the leadership of MCA. Finally, the role of the state is discussed more broadly, using the MCA as an example of how states can formally support the development of local food movements and markets as they mature and tackle complex problems in an integrated way. This case study differs from the other two in that the focus is not on a specific tortilla project. Maize, in the case of MCA, is integrated into a diversified strategy to support local farmers to access new markets, create new supply chain linkages, and shift toward agroecological production.

Michoacán's Agricultural Roots

Michoacán is situated in central western Mexico. It borders Jalisco and Guanajuato to the north, Querétaro to the northeast, Mexico and Guerrero to the east, and the Pacific Ocean to the west. The state is mountainous, crossed by the Sierra Madre and the Trans-Mexican Volcanic Belt. The state has seven different climactic regions, from coastal and tropical to dry mountain highland. Michoacán is made up of three large regions, divided into eighty-four municipalities. The cities of Morelia, Uruapan, and Zamora are where the majority of the population live (Mexico 2011).

There are 3,869,133 people in Michoacán and the indigenous population is approximately 254,319, or 7.1 percent of the population. Most indigenous people are located in the central Región Meseta-Cañada, Lacustre de Pátzcuaro and Zirahuén, and Ciénega de Zacapu. In the east of the state, in the municipalities of Ocampo, Anguangueo, Tuxpan, Hidalgo, Maravatío, and Zitácuaro, there are Mazahua, Otomi, and Purhepecha peoples. On the coast, Aquila, Chinicuila, and Coahuayana Nahuas peoples reside. The largest indigenous population is the Purhépecha, approximately 201,000 people in the central eastern portion of the state (Mexico 2011).

Driving into Michoacán from Mexico City to the east, the landscape is dry with rolling hills punctuated by the occasional impressive volcano rising from the milpas. The primary economic sector in the state is agriculture, making up 34 percent of employment. I immediately took notice of tree plantations that dotted the mountainsides—avocados. Michoacán is the largest producer of avocados in Mexico and in the world with a thriving export market. Later I would see huge avocado bodegas—packing plants—and visit several producer groups (see Riedemann and Huacuja 2007 and others for more on avocado production in Michoacán).

Agriculture across Michoacán represents approximately 1.5 million hectares, approximately 20 percent of the surface area of the country (Mexico 2011). After avocados, the principal crops grown are maize, sorghum, strawberries, peaches, wheat, lemon, pasture grasses (*pradera*), sugar cane, and mango, but over eighty-five agricultural products are grown across the state. Maize accounts for 43 percent of the crops harvested (Mexico 2011).

Organic agriculture has been growing quickly in Michoacán over the past decade, with organic avocado production leading the export market. The value of organic production in Michoacán rose 30 percent between 2008 and 2009, says Pablo Muñozledo Carrasco, the director of the project Michoacán Orgánico (Organic Consumers Association 2008). Although Michoacán does not have the largest land area under certified organic production, it may have the largest area under organic production.

Michoacán is the fourth largest maize producer on a national scale and grows 30 percent of Mexico's total maize crop area (Greenpeace 2011), producing 1.5 million tons of maize per year. (Michoacán Orgánico 2008). Most of the maize produced is consumed by the local market. Certified organic criollo maize production is also rising quickly, growing by 77 percent between 2008 and 2009 to satisfy the domestic demand for tortillas and tostadas (Michoacán Orgánico 2008).

Like the rest of Mexico, Michoacán agricultural producers suffered in the economic crisis over the 1980s. The economic crisis coincided with the strengthening of indigenous and civil society organizations and the rise of narco-trafficking. Both have intensified over the past few years. Narco-trafficking has become more prevalent in recent years with the Michoacán cartel a major center for production of methamphetamines and marijuana. Poverty and unemployment compel many people to be involved in the drug trade and many communities face daily threats and violence due to narco-trafficking (Hamilton 2011, 266). This economic crisis pervades the everyday life of the farmers, activists, and government employees I spoke with during my visits to Michoacán.

The origins of the Michoacán agricultural associations and indigenous organizations date back to the early twentieth century. Indigenous organizations and campesino movements have been strengthening over the past forty years due to agrarian and neoliberal policy reform and the economic crisis. Martínez and Jacaranda (2010, 73) describe how the demands of these movements, in particular the demands of indigenous movements, have evolved over the decades from a focus on communal rights to ethnic rights, and more recently regional rights. This shift is reflected in the approach taken by the MCA, as they focus working regionally and municipally on their agricultural development initiatives.

The Local Turn

At the root of MCA's work is the failure of global organic markets to economically sustain local organic farmers. In the early 1990s the rising global demand for organic products pulled farmers in Michoacán and beyond to transition to organic production and seek organic certification. As supply fought to keep up with consumer demand for organic products, prices rose. Smaller niche markets and distribution networks linked farmers to emerging markets in the United States, Canada, and Europe, and alternative, organic supply chains were developed. As the organic market grew, conventional supply chains adapted to include organic products, reproducing the structural issues—corporate concentration, larger scales of production and distribution, a greater number of players in the middle of the supply chain—that squeeze farmers on price at the farm level (see Dupuis 2002; Guthman 2004).

Michoacán campesinos, if they were able to access export markets, shifted from diversified production systems with maize at the center up until the 1960s, to specialized export-oriented agriculture. The markets became even more specialized for those who entered into the organic market, offering, at first, a better price than those offered by conventional buyers, and often with lower input costs. But as organic markets grew and became centralized, the farmers were squeezed on price. Jose Vilchis and Fulvio Gioanetto describe how many farmers are opting out of organic certification and choosing agroecological production systems to save on input costs and the costs of certification. While export markets are still important, direct local markets are also being developed.

The development of global standards for agriculture over the past decade (see Friedmann and McNair 2008) are also driving this focus on local markets because the costs of complying with these standards are too high for smaller farmers and the standards do not translate to local production practices.

In the late 1990s, Fulvio saw the organic market as an untapped opportunity for Michoacán farmers. He had recently moved to Mexico after working first as a researcher for the United Nations Environment Program and then as an organic farmer in Italy. In the mid-1990s, while traveling in Mexico as a consultant for the United Nations, he met Maria, who later became his wife. Fulvio moved to Nurío, a Purhepecha community outside of Uruapan. After working in Michoacán for several years, Fulvio started a chapter of the organic certification agency Bioagricoop in Mexico. Fulvio was a key contact for me in Michoacán, introducing me to many of the people involved in agricultural development and renewal in the state.

Bioagricoop thrived in the early 2000s, certifying many farmers. New labeling schemes, including organic and fair trade certifiers also moved into Michoacán during this time. As the global organic market began to grow, Fulvio became aware of the dwindling economic benefits to farmers. As a response to this he created Coyote Rojo in Michoacán, a bioregional, agroecological label developed for the local and national market (See Friedmann and McNair 2008). The Coyote Rojo label reflects a new trend in community certification, outlining a number of principles to guide production and distribution. In 2005 Fulvio became the manager of research and market development for MCA.

From Field to Table: Education for Agroecological Inputs, Enterprises, and Leadership

The MCA was founded in 2004 with a threefold mandate: first, to develop a commercial system that supports and consolidates the demand from agrifood businesses across the state; second, to promote and support the integration of production processes to improve the competitiveness of the sector; and third, to establish and strengthen internal food supply policies as a way of improving food security and services in the state. Although the agency does much to facilitate the export of Michoacán products, the recent priority of the MCA has been on regional and national supply chain development.

With agroecology and the development of local markets as the core focus, the MCA organizes a number of events, workshops and projects. These include workshops, trainings, and symposiums on criollo maize production and seed selection, organic production, organic aquaculture, organic kitchen gardens, and the development of regional labels. All of this work is done with the goal of developing local economic opportunities and jobs, as the largest social problem in Michoacán is unemployment.

PHOTO 5.1
Agroecological extension has been the cornerstone of the MCA's approach to supporting farmers.

Working with researchers from across Mexico and at various academic institutions, the MCA promotes the production of criollo maize varieties through the technical extension offered, as well as by organizing and participating in symposia on maize. In a small garden at the front of the MCA offices is a mipla with several different varieties of maize, as well as *teocinte*. The maize varieties were developed by Jose, Fulvio's colleague at MCA and the agency's director, who has been working to develop a perennial maize variety crossing various criollo varieties with *teocinte*, "reaching back" as he describes, "to maize's past to prepare for the future" (Gioanetto, pers. comm., May 9, 2008).

The MCA has a long-standing interest in the preservation and promotion of criollo maize. Jose has used his position with the MCA and his personal interest in maize to bring leading agronomists and anthropologists together to explore the possibility of developing a variety of perennial grain based on *teocinte*. The MCA works closely with three groups of criollo maize growers in Nurio, Pichatero indigenous groups, and the municipality of Tiquicheo, supporting seed selection and promoting the cultural uses of different varieties of maize in the community.

Jose was active in the January 2011 passing of a new law to manage and protect criollo maize as an important part of Michoacán's culinary inheritance (Ley de Fomento y Protección del Maíz Criollo como Patrimonio Alimentario de Michoacán). The law protects over eighteen of the fifty-nine criollo varieties of maize listed in Mexico, and bans the introduction or planting of genetically modified corn (*La Jornada Michoacán* 2011).

The production management systems promoted by the MCA deeply reflects agroecology principles of mimicking nature and draws on sophisticated organic technologies from Europe. These principles and technologies are applied to different scales of agriculture. On a walk through an organic blackberry field, bushes are carefully trellised and tended to maximize airflow and minimize mildew. Around the perimeter wild and weedy plants grow and are harvested to make foliar sprays that are applied by backpack sprayers. In backpack fumigators vegetable oil is combined with extracts from local plants to ensure that molds and mildews are kept at bay. Outside of Fulvio's Purhepecha village, north of Uruapan, he offers technical assistance to the farmers in their milpas. Again, wild and weedy plants are selected to make a foliar tea, this time applied by a hand spray bottle.

Fulvio, a botanist by training, has enlisted a network of academics and institutions to test the nutrient and micro nutrient values of these teas and foliar applications. The farmers connected to the MCA have access to very detailed analyses of their soils, crops, and pest problems. Some farmers have started ecological input businesses, selling organic plant extracts, fungal composts, and other biological inputs.

PHOTOS 5.2 AND 5.3
The MCA program supports farmers to create on-farm micro-enterprises making and selling ecological inputs.

Fulvio's conception of biodiversity reflects the traditional milpa. Although interested in maize agrobiodiversity, he clearly states that maize diversity is embedded within a rich agroecological system of plants that have important interrelationships. Working with farmers, agronomists, and scientists, he has set out a research agenda that includes identifying and solving regional agricultural challenges. The research results are disseminated in very practical workshops through the organic certificate program, showcasing research results, the products that farmers are developing, and practical techniques to produce the amendments.

MCA also supports the transition to organic production and provides funding for the three-year transition process. The first year of transition MCA provides the five hundred thousand pesos it costs for certification. The second year it provides 30 percent of the certification expenses. The MCA strategically focuses on promoting organic livestock production with traditional silvopastoralism techniques. MCA's programs do not support avocado production because it is resulting in mass deforestation throughout the state and they are committed to providing alternative crops. MCA also administers a revolving micro loan fund and provides certification and technical support for groups of small farmers interested in accessing the organic export market.

Several other programs are worth mentioning. The first is a state-supported universal food program, or a way to subsidize and distribute basic foods to vulnerable communities. The state of Michoacán purchases all the lentils grown in the state for this program. The second program is in the development phase and is called the "organic lunch" program. Through this program the state coordinates a school lunch program that uses local organic products—meat, dairy, grains/legumes, and fruit and vegetables—and is investing in the local logistics that will make this possible by linking regional farmers with nearby schools. The organic lunch program focuses on the social value of food and is a partnership between the school board, the teachers' union, farmers, and the MCA. The budget for this program is growing quickly as the state congress responds to increasing political will to support school food programs and tackle the childhood obesity crisis. The third program supports women to plant or improve their organic kitchen gardens. Finally, working with partners across the state, the MCA has also revived the *tianguis* (farmers market) and now have regular organic *tianguis* in Morelia and Uruapan, and occasional *tianguis* in other parts of the state.

Jose, the MCA's director, started working with the agency in 2008. He is appointed by the state government to this role and is a cabinet minister. Jose is deeply involved in Michoacán's social and indigenous movements. A coalition of social movement leaders meet regularly to discuss progress related to community development and *la lucha social*. This is separate from his work at MCA, but obviously connected. Jose is an "appointed" minister, so if there is

PHOTO 5.4
A store at the MCA's office sells local products to staff and visitors.

a change of government, he most likely won't keep his position. The program is therefore fragile in this sense, vis-à-vis the state, and there is uncertainty whether the work of MCA would continue under another political regime.

In this book, corn is used as a symbol for the commoditization of food and the corporate control of food production, processing, and consumption. In

BOX 5.2

When I was invited to speak to MCA's organic agriculture certificate program in 2009, I was the recently hired founding director of Sustain Ontario—The Alliance for Healthy Food and Farming. As the relationship with MCA evolved and we had the opportunity to understand each other's work, we were able to identify where each context had strengths and weaknesses related to food systems reform. The MCA and their partners were very interested in the multi-stakeholder, "big tent" model that Sustain Ontario was developing. I, on the other hand, was very interested in the agroecological training and enterprises that the MCA was facilitating. In 2011 the MCA was instrumental in developing the State Network for Agroecology and Regional Organic Production (*Red Estatal de Agroecología y Producciones Orgánicas Regionales*), formed to promote the consumption of healthy, organic, and local products, and develop common strategies to promote, distribute, and exchange experiences about organic and ecological production (Michoacán 2011b). The ideas and experience related to local food systems that are developed in one region can be the inspiration and a model for other regions.

this chapter, the work of MCA reflects the profound failure of the corporate, global food system to provide sustainable livelihoods for many agricultural farmers in Michoacán. Even shifting to take advantage of the organic niche has provided minimal economic benefits. Maize, on the other hand, is used as a symbol to describe agricultural and food practices grounded in practical farming knowledge, culinary traditions, and local economic exchanges. The work of MCA reveals the important role the state can play to promote diverse alternatives and support the transition to a local, sustainable food system.

Government Contributions to Local Food Networks

The work of one state agency in Michoacán, emerging from a unique political context and with unique leadership, illustrates how government resources can be leveraged to address a number of food issues simultaneously. Responding to economic, social, and dietary challenges, the agency works horizontally across government departments and with the community. The inability of export markets to sustain local farmers, the growing obesity epidemic, the chronic issues of access to healthy food, and the ecological degradation of agricultural land drives this work. These interrelated crises are examples of how contemporary agricultural and dietary transformations are permeating local places, and of how different actors are responding. In the case of MCA, a government agency is working with communities and social movements to acknowledge these crises and reclaiming local food systems. In the process they are regenerating local ecologies, cultures, and economies, and contesting neoliberal trade policies.

In this chapter, the challenges implicit with the dominant agricultural policy that props up export-driven agriculture and the co-optation of the organic niche by conventional supply chains have resulted in several simultaneous and interrelated processes of food system transformation. Transformations occur at the production level as farmers shift their practices to take advantage of export opportunities, and then again to take advantage of organic export opportunities. Corresponding social transformations occur as communities reorganize to aggregate products for export. One example of this is the Cupanda avocado growers cooperative, created in the 1950s to aggregate the avocado production in Taretan bioregion south of Uruapan. The coop is one of the largest exporters in Michoacán, yet it is seeking to diversify its marketing strategies, focusing on building market linkages into Mexico City and other Mexican urban centers in an attempt to shorten the supply chain and channel more of the consumer dollar to the farmers. The MCA is working with Cupanda on ecological production methods, as well as crop and market diversification.

The approach taken by the MCA and the strategies used to strengthen local food systems mirror the strategies of food movements around the world. In fact, the MCA is deeply connected to social movements in Michoacán and Mexico, as well as to the food movement through connections to groups developing fair trade initiatives and people developing shortened food supply chains within Mexico. Because of the MCA's position within government, this case study offers valuable lessons about how the state could better support local food system development.

These lessons can be grouped into three categories. First, the MCA is supporting farmers across the state to diversify their markets and create new market linkages. Although export markets are still encouraged, they are not considered an economic panacea but part of a diversified marketing strategy that includes local, regional, and national markets. A key element of this work on market diversification is a shift from supply chain development to value network development. This challenges conventional thinking about market pathways as linear supply chains, positing that the value created by agricultural products is enhanced through a network of relationships that includes all agricultural inputs, the community, the farmer, the distributor, the retailer, and the consumer—a broad range of actors that are seen to add value through their engagement in the network. An important aspect of this shift is that economic value for agricultural products is retained by the farmer as the network develops. The MCA works across Michoacán with farmers to build value networks that result in a higher share of the consumer dollar for the farmer. This is done through negotiating direct sales, developing local food branding or labeling schemes, creating new retail outlets for local products, and encouraging local food procurement in order to create high volume demand for local products. These efforts enable farmers to scale up their production through local markets and provide an expanded sense of economic opportunity that goes beyond the dominant market pathways available to farmers.

Second, the MCA offers lessons in how government agencies can support the transition to sustainable agricultural practices. The MCA supports farmers in their transition process from conventional production to organic production by paying for a declining portion of the organic certification costs over a three-year period. This investment distributes the risks associated with transition between the state and the farmer. The MCA also provides technical advice related to agroecological production methods, recognizing that organic certification is not necessarily appropriate for all farmers and that lower input costs can be gained through an agroecological approach. Most importantly, perhaps, is the MCA's investment in research and development related to agroecological production techniques and inputs made with local natural resources such as minerals, wild plants, and micro-organisms. By building relationships with academic scientists, the approaches promoted can be tested and enhanced. The MCA's research and development has spawned over one hundred new agroecological inputs enterprises, with farmers producing and selling biological inputs as part of their farm diversification strategies. The MCA has also started a municipal label program for the certification of local, agroecological products in three municipalities where the growers were facing challenges accessing the market. In order to promote new markets for farmers across the state, the MCA has opened five stores that sell local organic products (including meat, tortillas, and vegetables) in four cities.

Finally, the MCA provides lessons in the value of working horizontally across government divisions to embed agroecological and food systems thinking, leverage resources, and link agriculture to health and social well-being more generally. By taking a broad approach to the idea of market linkages and seeing local agriculture as part of the solution to the myriad of social, economic, environmental, and health crises challenging the state, new initiatives and partnerships have been created. Examples of this are the subsidized lentil program that purchases the harvest of four thousand hectares of local production to enhance healthy food access for vulnerable communities, the farm to school program, and linking with other social development efforts undertaken by the state. The practice and experience of "busting" traditional government silos challenges the entrenched institutional thinking that pervades most government departments. An important aspect of this cross-ministerial and departmental work is the ongoing articulation of the cultural and ecological importance of local food systems. The MCA is creating new organizational models that embed food and farming in appropriate ecological, cultural, social, and economic systems.

Conclusion

The case of MCA again provides a rich opportunity to think about the main themes developed in this book, providing a window into understanding key debates and tensions in the field of food studies: food security and food sovereignty, biodiversity and cultural diversity, culture and nature, the local and the global, markets and movements. The meeting of corn and maize offer new ways to understand these debates as fluid and mutually contested.

The work of MCA illustrates the complexity of reclaiming food from its commodity status in the global context of financial turmoil, food crises, and climate change. Place-based, the work of the MCA is grounded in the unique ecology, history, cultures, and economy of Michoacán. And yet the initiatives are not place-bound. They are networked locally, regionally, translocally, and transnationally through neoliberal policy, the circulation of agricultural products, and social movements.

Postscript

In 2012 there was an election in Michoacán that replaced the ruling leftist Party of the Democratic Revolution with the centrist Institutional Revolutionary Party. Both Jose and Fulvio left the MCA and their program has been dismantled. Their work has left an incredible legacy across the state. In 2011 Jose and Fulvio actively created the Network for Agroecology and Regional Organic Production that included growers, organic technical specialists, scientists, and consumers. This network continues to function, despite the withdrawl of the state from these activities. Trainings and workshops are still organized, bioregional and municipal labeling schemes continue to develop and gain interest from the public, and farmers continue to transition to organic, agroecological production methods with the support of well-trained and knowledgeable technicians and access to appropriate inputs.

6

Biocultural Agrifood Relations

WHAT CAN the Nuestro Maíz project, the Itanoní Tortillería, and the Michoacán Centre for Agribusiness tell us about food movements and markets? This chapter compares and contrasts the three initiatives in an effort to come to some conclusions about biocultural agrifood relations as they are constructed and reconstructed the world over.

The Vexing Neoliberal Context

From the introduction, *Corn Meets Maize* describes the impacts of the global regulatory framework and neoliberal policies as they impact particular localities. The role of the state in Mexico has shifted since 1982 when a structural adjustment program was imposed by international lending institutions. For rural campesinos, to characterize this shift broadly, state support for small-scale maize agriculture has been largely replaced by programs that encourage export-oriented production of fresh fruit and vegetables. This reality represents a harrowing context for the farmers and food networks described in the previous chapters.

The responses by civil society organizations to these neoliberal policy shifts has been varied and diverse. ANEC (Asociación Nacional de Empresas Comercializadoras de Productores del Campo, A.C.) struggles to cooperatively sell the maize of member farmers, eking out slightly higher prices for small- and medium-sized campesinos. The organization's solidarity markets attempt to buffer regions of the country negatively affected by difficult circumstances—

PHOTO 6.1
Neoliberal economic policy threatens Mexico's maize economy and agrobiodiversity.

drought, for example—when state and national level agriculture policy and programs do not have anything to offer. Despite the ongoing assault on Mexican campesinos and the direct attempt by federal bureaucrats to reduce their population (Esteva, pers. comm., June 24, 2008),[1] organizations like ANEC persist to support the work of small- and medium-scale farmers practically and ideologically. Advocating for policy changes and engaging with global movements like Via Campesina that are working to challenge and adapt to neoliberal policy shifts, ANEC is working to revalue the contribution of peasants not only in Mexico but around the world.

The Nuestro Maíz project illustrates how organizations like ANEC "interface with neoliberalism as both a material and discursive project" (Allen and Guthman 2006, 402). Patricia Allen and Julie Guthman, in their discussion of the Farm to School movement in the United States, describe how farm to school projects are "producing neoliberal forms and practices afresh" (2006, 401) in the ways that they adapt to public and private funding sources, devolve responsibility to the local level, and employ the rhetoric of neoliberal governmentality (Allen and Guthman 2006, 401, 410) through focusing on personal responsibility, consumerism, and choice (Allen and Guthman 2006, 410–411). This critique is useful for understanding how biocultural agrifood relations both contest and reflect neoliberalism, but perpetuates the alternative/conventional dualism this book attempts to transcend. ANEC, through the Nuestro Maíz project, responds to government programs to create rural enterprises and in the process encounters challenges related to the industrialization of the *nixtamal* process, ultimately resulting in the crop improvement program. As Allen and Guthman acknowledge, for the leaders of the Farm to School movement these are often tactical choices reflecting a narrow set of options. The unlikely alliances forged between movement leaders, farmers, and crop scientists points to the possibility for more democratic forms of hybridization (Kloppenburg 2004). The Nuestro Maíz project illustrates the possibilities and constraints of the neoliberal context, as well as the multifaceted, contradictory, and unstable nature of biocultural agrifood relations as they navigate this perilous terrain.

Many of the same issues are reflected in the Itanoní Tortillería case study. The Oaxaca context, however, sheds light on how uneven the neoliberal terrain is. Gustavo Esteva comments (pers. comm., June 24, 2008) that it is only because of the Oaxaca context that Itanoní is able to flourish. Oaxaca is alive with responses to neoliberalism—autonomous indigenous communities, alternative health practices, an independent university, a burgeoning nonprofit and civil society sector, and alternative economic initiatives based on the traditions of *mayordomo* and *tequio*,[2] including barter markets and local currencies. But within Oaxaca, neoliberalism has ravaged campesino farming systems, rural communities, and local economies and continues to do so.

The uneven impact of neoliberalism can be glimpsed through the descriptions of Hector and Luisa's and Chucho's experiences. Both farming families have felt the material impacts of shifting global and national policies. Because of their location in the Central Valleys of Oaxaca, Hector and Luisa have lost some of the cultural and ecological skills and knowledge related to maize production, influenced by the promise of improved seeds and technological packages to improve yield. Chucho, however, because of the isolation of his community and more challenging environmental conditions, has retained the technique of *maíz de cajete*. Yet his agricultural and farm management practices reflect the environmental movement in the region working on deforestation, water quality, and sustainable agriculture.

The Michoacán Centre for Agribusiness's activities offer an example of how the state can confront and address the uneven impacts of neoliberalism. By working with farmers across the state to build new market linkages based on value networks, by supporting the transition to agroecological production practices, and by working across government and institutional silos to embed food and farming as an essential social, cultural, and economic activity, the Michoacán Centre for Agribusiness (MCA) is actively building new biocultural agrifood relations to mitigate the impacts of neoliberal trade on communities that are dependent on agricultural trade, yet marginalized through global trade policies and practices.

The Nuestro Maíz, Itanoní Tortillería, and MCA reveal the way the "global permeates the local" (Braun 2006), illustrating how these food networks are place-based, but not place-bound. The initiatives are grounded in particular contexts and express some characteristics of the industrial food system. Nuestro Maíz, Itanoní Tortillería, and the MCA are engaged in the "active work of preservation" (Alvarez 2005), drawing on the cultural and culinary importance of maize in Mexico. They also develop innovative strategies to address the impacts of neoliberal policy shifts. The initiatives shed light on biocultural agrifood relations in the making, the imperfect process of culinary regeneration and local food innovation that can be wrought with challenges and contradictions.

These are cursory illustrations of the importance of context. Through examples like these, however, this book shows that context is indeed an extremely important aspect of whether local food networks flourish or fail. Examples of local food networks in different contexts illuminate the differing ecologies, cultures, and economies that shape biocultural agrifood relations.

The ongoing liberalization of maize trade in Mexico along with the GM (genetically modified) corn issue reinvigorated a broad number of social movement actors to address various facets of the food and agriculture system. ANEC's Nuestro Maíz, the Itanoní Tortillería, and the MCA are three examples of how diverse actors are developing place-based markets in response to neoliberalism. Harriet Friedmann suggests that social movements play (and have played historically) a critical role in regime crisis and formation (2005a, 231). These social movement responses "do not reject markets but rather seek to 'multiply niches' as an alternative strategy to the expansion of one set of products and procedures at the expense of all others (Fonte and Boccia 2004)" (Friedmann 2005a, 261). This book suggests that these translocally linked markets being developed by diverse actors in Mexico are part of a wider emergence of biocultural agrifood relations that model more democratic, ecological forms of growing, distributing, and consuming food.

Expanding Economies: From the Margins to the Center

It is a growing trend to seek out innovation in the margins of society. bell hooks conceptualizes this as a rupture of hegemonic power and an opportunity to shift our gaze from the dominant center to the borderlands (hooks 1994). Others follow suit—Morgan, Marsden, and Murdoch (2006) explore innovation and retro-innovation outside of the global corporate food production and distribution channels, but recognize how these food "systems" are not isolated, but interconnected as ideas, products, and technologies circulate. Virginia Nazarea (2005), in her study of seed savers from around the world, has perhaps the freshest approach to this focus on the margins. "Their marginality," states Nazarea,

> makes them effective and essential agents in the persistence of a counter memory that in turn undergirds conservation of plant genetic resources. Their life histories, their gardens, and their collections make us conscious of alternatives apart from the prescribed and the normalized; *a space on the side of the road* for reflection, transcendence, and transformation. Theirs is *an out-of-the-way place*, ripe with possibilities and promise. Their seeds encapsulate memory and constitute a narrative, a tangible link between past and present, between space and place. (2005, 153; italics in original)

This book draws on these ideas and their fluid notions of tradition, but challenges the dualistic metaphors of marginality and dominance. Instead, I explore a more dialectical understanding of local food networks. The work of Friedmann (2005a) and Pechlaner and Otero (2008) demonstrates how this activity "in the margins" is reflected in the quality audited supply chains of transnational agrifood corporations (Friedmann 2005a) and has, in some instances, the potential to shape state policy and corporate practices (Pechlaner and Otero 2008).

For Gibson-Graham, activities such as building local food networks represent ways to "perform the economy otherwise," to expand beyond notions of the economy defined solely by market relations and a focus on commodity production and capital accumulation. The three initiatives offer a glimpse into how these new economic possibilities and strategies are operationalized, and how they use opportunities and address challenges. I suggest that the projects focus more on the use-value than the exchange-value of maize. This shift in focus from production and capital accumulation to social reproduction can be seen in the way the initiatives articulate the importance of campesino livelihoods and the ecological aspects of production. Cultural and ecological values and ethics are embedded in their economic enterprises. Itanoní Tortillería embodies this in a deep, integrated way, as cooperative marketing structures are created and organic production methods implemented. The cooperative structure developed by Itanoní Tortillería is an example of how consumers and farmers can be "co-producers," as articulated by the Slow Food movement (Petrini 2005). Itanoní's educational work invites consumers to experience the taste of maize and learn about regional maize diversity and production, but also brings farmers together to share their production practices.

ANEC's mission to revalue campesino production and their efforts to introduce agroecological production methods illustrate how the discourse of food sovereignty is translated into practical projects on the ground by producer groups. The Nuestro Maíz project is attempting to build sustainable livelihoods for campesinos and create markets for their maize. Although ANEC is part of the global food sovereignty movement, the ideas espoused by this transnational movement are difficult to implement, resulting in a long transition period as ecological production methods are introduced and new economic models are developed.

At the MCA, the notion of value networks, in contrast to the conventional concept of value chains, offers an expanded way to think about social and economic relationships as new economic models are developed. The state invests in market linkages, diversification, ecological transition processes, as well as research and development in order to expand economic possibilities for farmers.

The work of ANEC's Nuestro Maíz project, the Itanoní Tortillería, and the MCA are examples of the heterogeneity of markets. Gibson-Graham's theorization of economic diversity (2006) is useful for understanding the potential and challenges faced by local food networks—challenges that are shaped by global economic and social processes.

The organizational structures of the initiatives represent models for social entrepreneurship, or *empresas sociales* (Villagomez 1999). Itanoní Tortillería is a horizontal organization that models a cooperative structure based on family and community. The horizontal nature of Itanoní is reflected in the informal networks and relationships with social movements and civil society organizations. Nuestro Maíz has a more vertical organization, with the project concept originating from ANEC's head office in Mexico City and applied across state contexts. This model transfer has not worked across the board and its success reflects the particular context in which it is applied. This more vertical structure is reflected in ANEC's formal relationship with the state and work on policy issues, as well as its participation in broader social movements and civil society organizations. The MCA is also creating new economic models through the creation of value networks, new agroecological enterprises, and horizontal, cross-ministerial, and departmental relationships based on real economic initiatives.

The three initiatives point toward the creative structures, alliances, and relationships, both horizontal and vertical, fostered by biocultural food networks. These represent the synergistic relationships between social movements and place-based, ecological markets within a translocal context. Innovative, synergistic relations are key elements of biocultural agrifood relations.

Verschoor describes the practices of maize movements and networks.

> These more ecological identities contrast sharply with neoliberal and reductionist narratives and prevent that cultural and agrobiodiversity be treated as contingent externalities of an inevitable process of technical and economic development. Instead, by putting uncertainty on the agenda, collective ecological identities demand renewed attention and political space. What was repressed through simplification, peasants, maize, genes, beliefs and so on, forcefully returns on the national and international agenda. (2007, 54)

The everyday practices and politics of the three initiatives reveal the dynamic tension between practical efforts to create sustainable and socially just economies and oppositional efforts based on protest and policy proposals. Stevenson (1998) describes the work of "builders" and "warriers" as essential to shifting toward agroecological food systems (see also Friedmann and McNair 2008). The three initiatives demonstrate that this work occurs simultaneously, through formal and informal political, scientific, and economic networks.

This book illuminates the importance of scale in constructing place-based alternatives. The small scale of Itanoní's activity and the broader scale of ANEC's Nuestro Maíz project present challenges to each initiative. Here, lessons from Brazil and Toronto can shed light on the issue of scale. In both

Brazil and Toronto, an institutional structure created through the state (in Toronto it is the municipal Toronto Food Policy Council and in Brazil the creation of councils at various levels—municipal, state, and federal) enables the replication and adaptation of ideas across contexts (Johnston and Baker 2005; Rocha 2001). The MCA also provides an example of how the state can support the creation of new institutional relations based on the active work of creating biocultural food systems. New relationships and alliances between unlikely players enable the proliferation of biocultural agrifood relations (see also Friedmann and McNair 2008).

The initiatives illustrate the potential for and limitations to galvanizing consumers to recognize and appreciate agrodiversity. Itanoní's focus on reclaiming regional culinary culture seeks to increase urban consciousness through the *taste* of agrodiversity. The work of Nuestro Maíz engages producer groups to close the production-consumption cycle by selling tortillas made from local maize in communities in the surrounding region. In the case of the MCA the value network recognizes the importance of maize as part of a diverse production system that provides for families, communities, and institutions. Consumer education emerged as an important focus for each project. Yet the initiatives lack institutional and policy support and resources that might help them scale up their activities (Johnston and Baker 2005). The initiatives are exploring ways to move beyond their current levels of activities and are seeking the institutional support necessary to do so.

Corn Meets Maize examines how initiatives such as Nuestro Maíz, Itanoní Tortillería, and the MCA encompass efforts to multiply local food networks while engaging with social movements working to challenge neoliberal policy at various levels. Translocal networks are created and re-created through solidarity work, transnational movements, and the circulation of ideas, technologies, and food. A close look at the shortened food supply chains offers perspective on building value-based economic enterprises embedded in cultural and ecological contexts. The initiatives also provide insight into the uneven terrain that must be navigated by these initiatives, and the contradictions and challenges their work faces as they focus on creating democratic processes and relationships over capital accumulation.

Maize's Agency

Throughout *Corn Meets Maize* I have illustrated how maize diversity interacts with people, institutions, technologies, and policies. In these interactions, maize expresses agency, shaping the initiatives. This follows the many ways maize has shaped agricultural landscapes and food economies historically. Maize diversity is a result of an intimate partnership with people, a partnership that has taken on many forms—agroecological and industrial. Following Bruce Braun, the initiatives illustrate how maize "adds something to the story of neoliberalism" (Braun 2006). This is an example of "the persistence of the organic" (Goodman and Watts 1997), as maize shapes the Itanoní Tortillería menu, the tortilla quality in the Nuestro Maíz *tortillerías*, and the seed selection and crop improvement efforts of the initiatives. At the MCA, the focus on agroecological inputs based on the surrounding diversity available, driven by the lower cost of these inputs for farmers and scientific research on their efficacy, also illustrates the way that the nonhuman elements of the food system shape the agricultural economy. These unanticipated consequences of the initiatives are interesting reflections of the ongoing reciprocal relationship between plants and people (Pollan 2002), and further illustrate the way that biocultural agrifood relations both contest and reflect neoliberalism.

Understanding the nature of maize's agency is difficult. How, exactly, does maize interact with farmers, landscapes, and food economies? Amado Ramírez Leyva brings this agency to life through a vivid example. At the *tortillería* one day we sit down to eat lunch together. On the table is a bowl of pozole for each of us, a *tetella espirituosa* (a triangular tortilla stuffed with the herb *oja de santa* and Oaxaca's famous cheese), and tortillas made from three different kinds of maize. Amado asks me to taste each item. The flavor of each maize variety is distinct. For the pozole and *tetella* it is also augmented by the flavors from other ingredients. Amado asks me which tortilla and dish I like best. My response, he explains, is maize asserting its agency. I choose to eat one of these dishes over another because I like the taste. If I were a farmer, I would choose the variety that I feel grows the best in my fields and meets the cultural and culinary needs of my family and community. And so maize expresses agency through this sensorial and practical process that honors embodied knowing. Amado states that this is the way in which both biodiversity and cultural diversity intersect with the identity of individuals and communities. Edit Antal also describes the intersection between diversity, power, and cultural practices in her work on the GM corn controversy in Mexico.

> The quality and purity of foodstuffs is not only a health issue, but also involves culture and survival. In the last analysis, the cultural norms and rules about food somehow reflect the distribution of power in society. Since in a rural society, eating is intimately linked to existential notions like fertility, health and illness, and therefore, birth and death, when peasant communities do not know what food is made of, it causes cultural confusion and engenders myths with the power to alter people's most basic notions. (Antal 2007a, 68)

The sensorial and spiritual qualities of maize are themes that run through the three initiatives. At Itanoní, the primary sensorial experience is taste. This is a taste of nostalgia and hope—nostalgia for the tortillas made by mothers and grandmothers on kitchen *comals* throughout Mexico, and hope for a renewed appreciation for maize diversity and the endurance of campesino cultures. At Nuestro Maíz, the sensorial experience is shaped by contrasting notions of quality as well as through the industrial production process tempered by the need to taste and smell the maize soaking with lime to see if the *nixtamal* process is complete. The MCA draws on the spiritual importance of maize in Mexico, and its centrality in religious and community festivals. It is through these expressions of nostalgia, authenticity, spirituality, quality, and taste that we better understand maize's agency in the initiatives. These examples illustrate how biocultural food networks extend agrifood relations beyond the human, integrating the multiple ecological and nonhuman aspects of food and agriculture.

The Challenge for Biocultural Agrifood Relations: Beyond Quality toward Connection

The work of projects like Itanoní, Nuestro Maíz, and the MCA, as well as the tireless documentation and categorization of maize landraces, only captures a small fraction of maize agrodiversity. The importance of these efforts lies in raising consciousness about the role of food and agriculture in biodiversity conservation.

Agricultural methods and food traditions have never been static, but have constantly been re-created to reflect local-global economies, ecologies, and cultures. Agricultural production techniques circulate globally; ideas about taste and quality shift. These shifts and circulations are evident in the description of initiatives in this book. Market forces and social movements have always and continue to shape agrodiversity. This book demonstrates how, in working toward agrodiversity conservation, it is important not to deny these shifts and circulations, but encourage their multiplicity. Diverse economies, cultures, and ecologies enable biodiversity.

The three initiatives illustrate the ways in which diversity is a multidimensional concept (Bellon and Berthaud 2006, 4) based on the interrelationship between plants, farmers, land, and eaters. Nuestro Maíz, Itanoní Tortillería, and the MCA reveal how notions of diversity shift over time. For the Nuestro Maíz project, diversity conservation was one goal of a value-added initiative. But ideas about maize diversity conservation were complicated by consumer preferences and the industrialized *nixtamal* technology. At Itanoní Tortillería, the owners are working to create an ethics and aesthetics of diversity, using their menu as an expression of maize diversity. The bodily act of eating engages a shift in consciousness about maize diversity. The MCA draws on diversity in a multifaceted way. The diversification of market opportunities is a key focus, as well as increasing the diversity of agroecological inputs used in production systems. Maize diversity, in this context, is advocated for through both the creation of markets for organic criollo maize, through valuing the ecological and cultural importance of agriculture, and also by passing laws that prohibit GM corn and promote agrodiversity. These intimate connections with maize can be understood through Toledo's concept of "historical synthesis," (Toledo 2001) as well as Morgan, Marsden, and Murdoch's (2006) notion of reaching back and bringing forward. Diversity reflects a dialectical process and cyclical relations. In summary, these projects demonstrate agricultural biodiversity as living and dynamic.

Central to this living diversity, revealed through the initiatives, is the work to recover, reclaim, and promote the skills of intercropping, ecological production, cultural and spiritual aspects of maize, and the taste and culinary use of multiple varieties. The projects are working on crop improvement through seed selection and, in the case of Nuestro Maíz, the introduction of hybrids. These examples stand in contrast to more static ideas about how agrodiversity can be conserved, preserved, and promoted such as categorization, documentation, seed banking, and even in situ conservation.

Quality-defined markets also perpetuate a static notion of food production and processing practices, an issue that is addressed in the work of the Slow Food Foundation's Ark of Taste project. Although there is some focus on quality, in documenting the work of artisanal food producers, connections between food, history, land, culture, and production processes are articulated. Maria Fonte (2006) describes how these connections and relationships are translated to distant consumers as a large retailer in Italy markets the products of some of Italy's Slow Food presidia.

In their study *The Practice of Everyday Life, Volume II: Living Cooking*, Michel de Certeau, Luce Giard, and Pierre Mayol (1998) explore the ways that currents of power and subversive creativity infuse everyday practices such as shopping, cooking, and eating. Culture, through these everyday acts, is cultivated. It is dynamic, changing, active. Gustavo Esteva and Michael Pollan suggest that food culture is transformational. Their critique of the nutritional sciences and formulaic ways of counting calories and defining health is countered by a simple meal, rooted in the local ecology, culture, and economy of a particular place (Esteva 1996; Pollan 2006). These ideas about agriculture and eating are brought together by Itanoní. Itanoní's *comal* can be thought of as a bridging metaphor. The center of the restaurant, the *comal* reflects the culinary traditions of Oaxaca, offering a place from which the *guisos* and maize—the identity and diversity of the people and places where they were grown—can be prepared and tasted.

The initiatives provide expressions of biocultural agrifood relations—intimate connections between people and plants, agricultural production and culinary culture, and the practical work of building local food networks and oppositional political movements. These connections result, for example, in the Nuestro Maíz brochure proclaiming their tortillas as "100% maíz Mexicano" and in the menu at Itanoní, from which consumers can select dishes re-created from the "spaces of memory and life." As was so eloquently stated by Fulvio Gioanetto from the MCA, this represents "the connecting threads between the global imagination and local landscapes" (pers. comm., May 9, 2008).

Place-based, but Not Place Bound: Contestation in the Neoliberal Context

Corn Meets Maize examines how neoliberal policy shifts and new agricultural biotechnologies in Mexico galvanize social movements to experiment with new ways of marketing maize. These market-based strategies have multiple goals including rural development and agrodiversity conservation. A close examination of three such initiatives—ANEC's Nuestro Maíz project, the Itanoní Tortillería, and the MCA—reveal their challenges and potential. The projects reflect and contest neoliberalism, government policy, and particular local contexts. They draw on old forms of political organizing and create new forms of politics and organizational structures to support their ideas. I argue in this book that the strength of biocultural food networks is that they are relational—connected to particular cultures, ecologies, and economies, as well as translocally to other markets and social movements.

The initiatives I examine are illustrative of biocultural agrifood relations, a concept that I developed to capture the proliferating work of food movements to create food supply chains that are grounded in democratic processes and ecological principles. Mexico is an important place to examine these biocultural relations. A unique constellation of agricultural trade and biotechnology policies, the historical and cultural importance of maize, and strong social movements point toward how biocultural agrifood relations can potentially shape national policy and perhaps emerging food regimes (Pechlaner and Otero 2008; Friedmann 2005a).

Corn Meets Maize uses the notion of biocultural agrifood relations to explore the relational aspects of translocal food networks. The shortened maize supply chains studied are creating democratic processes and creative organizational structures that (re)embed food and agriculture within particular cultures, economies, and ecologies. Yet these very local, place-based initiatives are connected relationally through markets, personal networks, and social movements, redefining key relationships. This can be seen in the crop improvement programs of Nuestro Maíz, Itanoní Tortillería, and the MCA, where farmers, scientists, and activists come together to address agrodiversity conservation. This book reveals how notions of quality are contested in these relationships, as well as how farmers and scientists are working to democratize the process of improving yield, the taste of tortillas, and market development (Kloppenburg 2004; Friedmann 2003).

In the initiatives, close attention to maize as an *actor* illustrates how place is shaped by nonhuman agency. This book addresses several tensions in the political ecology and food studies literature—the tension between nature/culture, the global/local and markets and social movements—illustrating how these tensions are infused by an expanded notion of democracy, one that includes the agency of nonhuman actors. In the food studies literature, this discussion of agency expands on the notion of the "persistence of the organic" through food chains and movements (Goodman and Watts 1997). This book demonstrates how biocultural agrifood relations are place-based, but not place-bound, drawing on metaphors of flow and flux to understand these relational networks as "in the making" (Sundberg 2004).

Corn Meets Maize focuses on linking the everyday practice of growing, exchanging, and cooking maize in specific cultural, economic, and ecological contexts to broader social movements addressing food and agriculture in Mexico and beyond. The notion of biocultural agrifood relations centers the ecological and cultural elements of these linkages, while focusing on the relational aspects of the creative strategies, networks, and local economies forged by the maize initiatives studied.

The Mexican examples illustrate how initiatives such as Nuestro Maíz, Itanoní Tortillería, and the MCA are proliferating niche markets and developing creative strategies to embed their shortened food supply chains in democratic processes and local ecologies. This represents a bringing together of politics and ecology that has unanticipated and surprising results. As noted previously, for Nuestro Maíz the new relationships between agronomists, farmers, and ANEC demands a more democratic approach to crop improvement (Kloppenburg 2004). Solidarity markets link producer groups translocally in a policy environment where the government has shifted from supporting campesino production to export-oriented agriculture. Quality emerges as a contested concept, one that simultaneously reflects the local maize crop, the *nixtamal* process, and the ways that diets and tastes are changing. By focusing on the *relational* aspects of these shifts, contradictions, and linkages, the notion of biocultural agrifood relations contributes to the food studies literature that moves beyond ideas of alternative/local and conventional/global and "quality" food systems (Allen et al. 2003; Mansfield 2003; Stassart, Whatmore, and Renting 2003; Bryant and Goodman 2004; DuPuis and Goodman 2005). By actively exploring how the maize-human relationship continues to shape diversity, this book reveals the relational aspects of the nature-culture tension. My focus on local, place-based initiatives illustrates how global (and national) markets and policies permeate particular contexts. These supply chains are connected through organizations and individuals to transnational social movements such as Via Campesina and Slow Food.

Two examples serve, in conclusion, to illustrate what is biocultural about food networks and provide an opportunity to return to the three central tensions identified in earlier chapters: nature-culture, global-local, and markets-movements.

Sopa de Milpa

At Itanoní Tortillería, I was particularly delighted by one dish. The *sopa de milpa* is a simple soup made with squash blossoms, leaves, and tendrils; chunks of corn on the cob; and the weedy greens and herbs so commonly found growing in and around the maize fields in Mesoamerica. The aromatic herbs create a light and fragrant broth, the greens are flavorful, and the corn provides hearty satisfaction. In the Oaxacan markets, the ingredients are sold mixed together in plastic bags, inviting the cook to just throw the contents in a pot and simmer.

Several years after this initial taste of the *sopa de milpa*, I was in my garden in Toronto. The pumpkin my son insisted on planting in our very small backyard was sprawling over the garden paths. Remembering the *sopa de milpa* and just having bought the first corn of the season at the farmer's market, I cut the squash blossoms and tendrils as well as some of the herbs and edible weeds growing in the garden. I simmered some onion and then threw the ingredients into a soup pot. The result was a taste that transported me back to Oaxaca.

The *sopa de milpa* is a metaphor for biocultural agrifood relations, reflecting the culinary potential of polycrop production, the ecological synergies between the crops, and the multidimensional aspects of diversity that include the interrelationships between land, plants, farmers, and eaters. My garden and kitchen experiments speak to the global circulation of food ideas and tastes. And yet the biodiversity in my garden is completely different from the diversity found in a Mexican milpa. Although the taste conjures memories of the *sopa de milpa* I ate in Oaxaca, it is also distinct because of the variety of squash I grew, the different kind of maize available in Ontario, and the herbs in my garden. The *sopa de milpa* embodies the interrelationship between nature and culture explored through this book.

Biocultural agrifood relations are, as I have illustrated, place-based and context specific. But this example of transplanted culinary inspiration illustrates how the global and local are interconnected through networks and exchanges. Another story illuminates this point. In my Toronto neighborhood tucked behind a Loblaws supermarket and some social housing buildings is a milpa. Michael Sacco, the owner of ChocoSol, a company that has created horizontal trade networks between Toronto, Oaxaca, and Chiapas, planted the milpa with seeds brought from Mexico. The milpa is thriving: tall maize plants are interspersed with sprawling squashes, beans, herbs, and edible weeds. It embodies the global-local circulation of seeds, agricultural practices, and culinary traditions.

Via Campesina and Slow Food

A second example extends the markets-movements discussion explored in this book. Via Campesina and Slow Food reflect the diversity of strategies used by transnational social movements as they organize and coalesce to form biocultural food relations. Both movements are actively creating new markets, but their strategies differ. Via Campesina is a more traditionally oppositional effort, challenging neoliberal policy by focusing on global bodies such as the World Trade Organization, articulating the impact of these policies on the peasants that make up the organization's transnational constituency, and proposing new rules for valuing small-scale agroecological production. Philip McMichael describes how Via Campesina has shifted the ontological foundation of agrarian movements by "combining a politics of circulation with a politics of production" (2008, 213). The movement moves beyond articulating a critique of agriculture in the twenty-first century to revalue diverse peasant production systems grounded in social and ecological sustainability (210). "In the current agrarian question posed by the food sovereignty movement, food embodies a broader set of relations, becoming a window on the social, demographic and ecological catastrophe of neoliberalism" (211). McMichael further notes,

> A grounded ecological perspective offers a range of values concerning the multifunctional and epistemic contributions of agriculture to humans and nature alike. This is the ontological break that informs the food sovereignty movement and its advocacy of revaluing small farming. . . . Building such a political space is

> a component of the food sovereignty movement's consolidation of an ontological detour that rejects the grand narrative of modernity, industrialism and proletarianization as an unfulfilled dream/palpable nightmare, and affirms an alternative, historically-grounded narrative sensitive to place and value incommensurability, as concrete universals, rather than the abstract logic that justifies accumulation by dispossession. This is an ontology that offers a politics of voice and struggle *on* the land, in addition to struggle *for* land (Flavio and Sanchez 2000), thereby politicizing the social-ecology of poverty relations. (214; italics in original)

The Slow Food movement also has a critique of neoliberal policy, in particular the global corporate food supply and distribution system, and has developed a way to revalue traditional, place-based production methods, linking "food communities" through their Terra Madre network and the Ark of Taste project. Central for Carlos Petrini, founder of Slow Food, are the principles of "good, clean and fair" (2005). Slow Food's work has put forth the concept of "co-producers" of food. "We must revive the system of local adaptation, we must relocate, and we must be aware that food is a network of co-production, where knowledge is shared and methods of production are sustainable. Food must be seen as a network, which thus forms another network, consisting of food communities" (2005, 174). Bringing together co-producers from across the globe for the biannual Terra Madre meeting builds on Petrini's belief in the educational value of travel.

> No system of transmitting knowledge within the network can ever replace direct contact. As gastronomic science teaches us, the exercise of one's own sensoriality, trying, tasting, contact with producers and with chefs, is the best way of learning how to interpret reality. To achieve this form of learning, it is necessary to *move*, to meet people, to experience other territories and other tables. If we apply this conviction to the network, it is vital to guarantee the circulation within it of people, from one side of the globe to the other, without distinction and without restriction. The *right to travel* becomes fundamental, a premise on which to base cultural growth and the self-nourishment of the network of gastronomes. (Petrini 2005, 241)

The functions of the network are described as communicating, recording, trading, and traveling. For the Slow Food movement, the market represents a way to achieve some of these functions. The market, however, is embedded within cultures and ecologies from which spring unique food traditions and products. In an article elaborating on the concept of food as network, Carlo Bogliotti describes the network as a highly democratic tool (2006, 70) to circulate information, facilitate markets, and enable cultural and culinary exchange. Central is the idea that gastronomic and agricultural knowledge circulates, is exchanged, and is in dialogue with modern science (Bogliotti 2006, 65).

Maria Fonte details an agreement between Slow Food and Coop Italia, a large retailer in Italy, to sell "typical" products made by Slow Food presidia. This represents "local production for distant consumers" (Fonte 2006) and illustrates the complexities involved in scaling up and out the food network for "reflexive markets for quality food" (Fonte 2006, 33). Friedmann and McNair also document food networks embedded within cultural and ecological contexts, using the example of Coyote Rojo described in the previous chapter and Toronto's Local Food Plus initiative linking local farmers with institutional buyers through a label that addresses ecological and social issues along the production-consumption chain (Friedmann and McNair 2008).

These examples of recent research on food networks speak to the challenges of building new ways to grow and eat within the neoliberal context, as well as an emerging "new constellation of agrifood relations and practices" (Friedmann and McNair 2008, 429). These initiatives, as this book demonstrates, are permeated by the neoliberal context in which they are embedded and reflect the challenges of the One No, Many Yeses strategy of the Zapatista movement (Esteva 2005), and a "unity in diversity" (McMichael 2008, 207).

I offer these two examples—the *sopa de milpa* and an elaboration of Via Campesina and Slow Food—as a way to illuminate the ways the three tensions—culture-nature, global-local, and markets-movements—remain in dialectical relationship with each other.

Conclusion

The sensorial experiences associated with growing, processing, and eating maize perhaps best embody the notion of biocultural agrifood relations. Ever changing and shifting, reflecting broader economic, cultural, and political processes, grounded in particular places but simultaneously linked translocally through social movements and the global economy, biocultural agrifood relations are places where cultural and agricultural diversity intersect. Through the biocultural lens the conditions that support diversity are revealed to be context specific yet embedded in the global circulation of ideas, products, and technologies.

The notion of biocultural agrifood relations offers an analytic tool through which to examine the proliferation of local food networks. In the case of Nuestro Maíz, a biocultural analysis offers a way to understand how neoliberal policy infuses the practices and politics of social movements working to create alternative economic avenues for food producers and rural communities. For Itanoní Tortillería, a biocultural analysis enables us to see the importance of creating an ethics and aesthetics of diversity. The initiatives are illustrative of the creative strategies and democratic processes used by activists and entrepreneurs to bridge markets and social movements.

In chapter 1 I reviewed the food security–food sovereignty and biodiversity-biotechnology debates. Maize movements in Mexico address the material conditions of food security, challenging neoliberal policy and asserting the value of campesino production systems. These movements challenge static definitions of maize diversity and acknowledge the dynamic and intimate ties between maize, people, and agricultural, cultural, and culinary practices. The initiatives were contextualized in chapter 2, where I examined the factors affecting maize diversity to illustrate the historical and cultural importance of maize in Mexico. I then reviewed policy shifts since 1982 to document how deeply the Mexican agricultural sector and maize production have been affected by neoliberal policy. The three case study chapters provided insight into the activities of three initiatives that are connected to transnational social movements and also actively building and experimenting with local food networks. The initiatives demonstrate the potential and challenges for biocultural agrifood relations, as well as the creative strategies used by such initiatives to build local markets, educate consumers, and link to broader social movements. The stories describe the complexity of creating shortened food supply chains—local food networks—in the neoliberal context, and embedding the economy within cultural and ecological values and ethics. Tracing the circulation of agricultural policies, production, and processing methods of maize and tortillas reveals how diversity is perpetuated by local food networks in unexpected and dynamic ways.

Postscript: Grounded Translocal Exchanges

Throughout my connection to Mexico since 1997 there have been several memorable opportunities to host Mexicans in Toronto and introduce them to the many interesting food projects in the city. These visits have been *exchanges* between practitioners and builders of local food networks and have provided the opportunity to share ideas and experience across very different contexts. Two visits stand out.

Isabel

The first was in October 2001 when Isabel, a migrant agricultural worker from Miacatlán, Morelos, was working at a berry farm in southern Ontario and extended her time in Ontario to work with me in the Urban Agriculture Program at FoodShare.[3] Side by side we shared the daily garden and greenhouse work. Isabel was used to hard work and we were able to quickly finish the tasks set out for the day, allowing time for conversation and tours of community food projects.

Isabel, now close to sixty years old, has been coming to Ontario for almost twenty years. When she visited me in 2001 she had never been to Toronto. Upon seeing the Good Food Box packed for delivery to communities across Toronto, Isabel remarked that she was going to take this idea home and implement it in her community. Her insightful observations about our project's sprouting and greenhouse production and ideas about how to outreach to Toronto's Latino community helped to shape the program. There are many benefits and outcomes from such an exchange. Beyond the inspiration drawn from the work Isabel participated in over her week-long visit to Toronto, she provided me and others with a glimpse into the life of agricultural workers in Ontario, the backbone of our local agricultural economy.

Fulvio and Amado

In May 2008 Amber McNair, a doctoral student at the University of Toronto, and I hosted Fulvio Gioanetto, from the MCA, and Amado Ramírez Leyva, the owner of Itanoní Tortillería in Oaxaca. Together, we spent ten days touring farms and food projects, developing the seed of an idea for a translocal project linking the educational and marketing expertise we observed in Ontario with the production expertise observed in Mexico. At a public presentation at the Big Carrot in Toronto, Fulvio and Amado profoundly articulated the interconnections between culture, ecology, and economy and the importance of translocal understanding. The story of their respective projects deepened the audience's analysis of the global food crisis as a structural and ecological crisis. Grounded in the local context, global exchanges are even more rich and relevant.

These two examples of North-South exchange illustrate the potential of translocal biocultural agrifood relations. Breaking the pattern of northern experts going to the south to "fix" problems, the relational aspect of these exchanges, and their grounding in local ecologies and cultures disrupt asymmetrical power relations to connect practitioners of biocultural food networks who are deeply knowledgeable about their particular places. Dialogue and information exchange, generating exciting ideas based on strengths of each context and working together on cross-border initiatives, allow for horizontal linkages across local contexts.

Touring food projects in Ontario with Mexican colleagues was a powerful reminder of my own history and role in building the local food economy in Ontario. Since that tour I have delved back into the practical work of building what

I now understand as biocultural agrifood relations. For me this represents a balance between theory and practice and a way to ground this book in my experience working on these complex and contradictory issues in my own context.

Notes

1. Esteva described, during a course he taught at York University in the spring of 2008, how Mexican government bureaucrats articulate a clear goal to reduce the number of campesino farmers over the next decade. The combination of policy shifts away from programs that support small-scale subsistence farmers and open maize markets make it increasingly difficult for campesinos to continue farming.
2. *Mayordomo* and *tequio* are social institutions in Mexico, important for the day-to-day functioning of the community. *Mayordomo* is an unpaid position to organize community festivals. *Tequio* is unpaid, obligatory work for the benefit of the community.
3. I got to know Isabel through my work with Deborah Barndt on the Tomasita project.

References

Acuña Rodarte, O. 2003. Toward an Equitable, Inclusive, and Sustainable Agriculture: Mexico's Basic Grains Producers Unite. In *Confronting Globalization: Economic Integration and Popular Resistance in Mexico*, ed. T. Wise, H. Salazar, and L. Carlsen, 129–148. Bloomfield, CT: Kumarian Press.

Allen, P. 1991. Reweaving the Food Security Safety Net: Mediating Entitlement and Entrepreneurship. *Agriculture and Human Values* 16 (2): 117–129.

Allen, P. 2004. *Together at the Table: Sustainability and Sustenance in the American Agrifood System.* University Park: Pennsylvania State University Press.

Allen, P., M. FitzSimmons, M. Goodman, and K. Warner. 2003. Shifting Plates in the Agrifood Landscape: The Tectonics of Alternative Agrifood Initiatives in California. *Journal of Rural Studies* 19:61–75.

Allen, P., and J. Guthman. 2006. From "Old School" to "Farm-to-School": Neoliberalization from the Ground Up. *Agriculture and Human Values* 23 (4): 401–415.

Allen, P., D. Van Dusen, J. Lundy, and S. Gliessman. 1991. Integrating Social, Environmental, and Economic Issues in Sustainable Agriculture. *American Journal of Alternative Agriculture* 6 (1): 34–39.

Altieri, M. 2000. Agroecology: Principles and Strategies for Designing Sustainable Farming Systems. *Agroecology in Action* (online). Available from http://agroeco.org/ (accessed September 18, 2008).

Altieri, M. A., and S. B. Hecht, eds. 1990. *Agroecology and Small Farm Development.* Boca Raton, FL: CRC Press.

Alvarez, R. 2005. *Mangos, Chiles and Truckers: The Business of Transnationalism.* Minneapolis: University of Minnesota Press.

Amin, A. 2002. Spatialities of Globalization. *Environment and Planning A* 34:385–399.

Anderson, M., and J. Cook. 1999. Community Food Security: Practice in Need of Theory? *Agriculture and Human Values* 16 (2): 141–150.

Antal, E. 2004. Who Should Tell Me What to Eat? *Ecology, Voices of Mexico* 68:113–117.

Antal, E. 2007a. Between Global Regimes and Local Demands: The Regulation. In *Maize and Biosecurity in Mexico*, ed. E. Antal, L. Baker, and G. Verschoor, 11–32. Cuadernos de CEDLA, 22. Amsterdam: Centre for Latin American Research and Documentation.

Antal, E. 2007b. Debate on Biosecurity in Mexico in the North American Context. Paper presented at the Latin American Studies Association Meeting, Montreal, September 5–8.

Antal, E., L. Baker, and G. Verschoor, eds. 2007. *Maize and Biosecurity in Mexico: Debate and Practice*. Cuadernos de CEDLA, 22. Amsterdam: Centre for Latin American Research and Documentation.

Appendini, K. 1996. *Changing Agrarian Institutions: Interpreting the Contradictions*. CERLAC Working Paper Series. Toronto: York University.

Appendini, K. 2001. *De la Milpa a los Tortibonos: La Restructuracion de la Politica Alimentaria en Mexico*. Mexico City: El Colegio de Mexico, Instituto de Investigaciones de las Naciones Unidas para el Desarrollo Social.

Babbie, E. 1998. *The Practice of Social Research*. Belmont, CA: Wadsworth.

Badstue, L. B. 2006. Smallholder Seed Practices: Maize Seed Management in the Central Valleys of Oaxaca, Mexico. PhD diss., Wageningen University.

Baker, L. 1999. A Different Tomato: Creating Vernacular Foodscapes. In *Women Working the NAFTA Food Chain: Women, Food and Globalization*, ed. D. Barndt, 249–260. Toronto: Second Story Press.

Baker, L. 2002. *Seeds of Our City: Case Studies of Eight Community Gardens in Toronto*. Toronto: FoodShare.

Baker, L. 2004. Tending Cultural Landscapes and Food Citizenship in Toronto's Community Gardens. *Geographical Review* 94 (3): 305–325.

Baker, L. 2007. Regional Maize Marketing Initiatives. In *Maize and Biosecurity in Mexico*, ed. E. Antal, L. Baker, and G. Verschoor, 55–80. Cuadernos de CEDLA, 22. Amsterdam: Centre for Latin American Research and Documentation.

Baker, L. 2008. Local Food Networks and Maize Agrodiversity Conservation: Two Case Studies from Mexico. *Local Environment, Special Issue on Local Food Systems and Direct Marketing* 13 (2): 1–16.

Baker, L., and J. Huh. 2003. Rich Harvest. *Alternatives Journal* 29 (1): 21–25.

Barabas, A. M., M. A. Bartolome, and B. Maldonado. 2003. *Los Pueblos Indigenas de Oaxaca: Atlas Etnografico*. Mexico City: Instituto Nacional de Antropologia e Historia.

Barkin, D. 2002. La Soberania Alimantario: El Quehacer del Campesinado Mexicano. *Estudios Agrarios*, VII Premio Estudios Agrarios.

Barkin, D. 2003. El Maiz y la Economia. In *Sin Maiz no Hay Pais*, ed. G. Esteva and C. Marielle, 155–175. Mexico City, Museo Nacional de Culturas Populares.

Barndt, D., ed. 1999. *Women Working the NAFTA Food Chain: Women Food and Globalization*. Toronto: Second Story Press.

Barndt, D. 2002. *Tangled Routes: Women, Work and Globalization on the Tomato Trail*. Aurora, Ontario: Garamond Press.

Bartra, A. 2005. Milpas Airadas: Toward Food and Labour Sovereignty. In *Mexico in Transition: Neoliberal Globalism, the State and Civil Society*, ed. G. Otero, London: Zed Books.

Batalla, G. B. 1996. *Mexico Profundo: Reclaiming a Civilization*. Austin: University of Texas Press.

Bellon, M., and J. Berthaud. 2006. Traditional Mexican Agricultural Systems and the Potential Impacts of Transgenic Varieties on Corn Diversity. *Agriculture and Human Values* 23:3–14.

Berry, W. 1990. The Pleasures of Eating. In *Our Sustainable Table*, ed. Robert Clark. San Francisco, CA: North Point Press. Originally published in the *Journal of Gastronomy* 5 (2).

Bogliotti, C. 2006. The Future of Food. *Slow: The Magazine of the Slow Food Movement* 53:64–71.

Bordi, I. V. 2006. The "Authentic" Taco and Peasant Women: Nostalgic Consumption in the Era of Globalization. *Culture, Agriculture, Food and Environment* 28 (2):97–107.

Born, B. and M. Purcell. 2006. Avoiding the Local Trap. *Journal of Planning Education and Research* 26:195–207.

Braun, B. 2004. Querying Posthumanism. *Geoforum* 35 (3): 269–273.

Braun, B. 2005. Environmental Issues: Writing a More-than-Human Urban Geography. *Progress in Human Geography* 29 (5): 635–650.

Braun, B. 2006. Environmental Issues: Global Natures in the Space of Assemblage. *Progress in Human Geography* 30 (5): 644–654.

Braun, B., and N. Castree. 1998. *Remaking Reality: Nature at the Millenium*. London: Routledge.

Bressani, R. 1990. Chemistry, Technology, and Nutritive Value of Maize Tortillas. *Food Reviews International* 6 (2): 225–264.

Brush, S. B. 2004. *Farmer's Bounty: Locating Crop Diversity in the Contemporary World*. New Haven, CT: Yale University Press.

Brush, S. B., and H. Perales. 2007. A Maize Landscape: Ethnicity and Agro-biodiversity in Chiapas Mexico. *Agriculture, Ecosystems and Environment* 121 (3): 211–221.

Bryant, R. L., and M. K. Goodman. 2004. Consuming Narratives: The Political Ecology of "Alternative" Consumption. *Transactions of the Institute of British Geographers* 29 (3): 344–366.

Cameron, J., and K. Gibson. 2005. Participatory Action Research in a Poststructuralist Vein. *Geoforum* 36 (3): 315–331.

CECCAM (Centro de Estudios para el Cambio en el Campo Mexicano) et al. 2003. Contaminacion Transgenica del Maiz en Mexico: Muchos mas Graze. Collective Press Release, October 9, Mexico City, Mexico. http://www.etcgroup.org/node/148 (accessed April 7, 2007).

Centre for Studies in Food Security. n.d. Food Security Defined. www.ryerson.ca/foodsecurity/ (accessed June 30, 2007).

Centro de Investigaciones Economicas, Sociales y Technologicas de la Agroindustria y la Agricultura Mundial. 2006. Mexican Network of Tianguis and Organic Markets. http://ciestaam.edu.mx/ (accessed July 11, 2006).

Cevallos, D. 2006. Environment: Mexico Shuts the Door on GM Maize. Interpress Service News Agency, Oct 28. ipsnews.net/news.asp?idnews=35280 (accessed April 22, 2007).

Chambers, K., and J. Momsen. 2007. From the Kitchen and the Field: Gender and Maize Diversity in the Bajío of Mexico. Paper presented at the Latin American Studies Association Meeting, Montreal, September 5–8.

Chavez, M. 2002. The Transformation of Mexican Retailing with NAFTA. *Development Policy Review* 20 (4): 503–513.

CNIPMT (Cámara Nacional de la Industria de la Producción de la Masa y la Tortilla). 2008. Estadísticas Nacionales. www.cnmaiz.org.mx/estadisticas_nacionales.php (accessed October 21, 2008).

Coleman, W. D. 2003. *Globality and Transnational Policy-Making in Agriculture: Complexity, Contradictions, and Conflict.* Hamilton: Institute on Globalization and the Human Condition.

Commission for Environmental Cooperation (CEC). 2004. Maize and Biodiversity: The Effects of Transgenic Maize in Mexico. Montreal Commission for Environmental Cooperation.

Cook, I., and P. Crang. 1996. The World on a Plate: Culinary Culture, Displacement and Geographical Knowledges. *Journal of Material Culture* 1 (2): 131–154.

Cook, I., et al. 2006. Geographies of Food: Following. *Progress in Human Geography* 30:655–666.

Conway, J. 2004. *Identity, Place, Knowledge: Social Movements Contesting Globalization*. Toronto: Fernwood Publishing.

Cornelious, W. A., and D. Myhre, eds. 1998. *The Transformation of Rural Mexico: Reforming the Ejido Sector*. U.S.-Mexico Contemporary Perspectives Series (USA), no. 12. San Diego: Center for U.S.-Mexican Studies.

Council on Hemispheric Affairs. 2008. Calderon, NAFTA and Mexico's Campesinos in 2008. April 1. www.coha.org/2008/04/01/calderon-nafta-and-mexico%E2%80%99s-campesinos-in-2008/ (accessed April 18, 2008).

Crone, I. 2007. Dollars and Sense: The Corn Conundrum. *Columbia Political Review*. www.cpreview.org/pdfs/200712.pdf (accessed July 2, 2008).

Cummings, C. H. 2002. Risking Corn, Risking Culture. *World Watch* 15 (6): 8–19.

Dahl-Bredine, Phil. 2006. Case Study: Food Sovereignty in the Mixteca Alta. In *Agroecology and the Struggle for Food Sovereignty in the Americas*, ed. A. Cohn, J. Cook, M. Fernández, R. Reider, and C. Steward, 63–66. London: International Institute for Environment and Development.

Dalgaard, T., N. Halberg, and J. Fenger. 2002. Can Organic Farming Help to Reduce National Energy Consumption and Emissions of Greenhouse Gasses in Denmark? In *Economics of Sustainable Energy in Agriculture*, ed. E. C. van Lerland and A. O. Lansink, 191–204. Dordrecht, The Netherlands: Kluwer Academic Publishers.

de Certeau, M., L. Giard, and P. Mayol. 1998. *The Practice of Everyday Life, Volume II: Living Cooking*, trans. T. J. Tomasik. Minneapolis: University of Minnesota Press.

DeLind, L. 2002. Place, Work, and Civic Agriculture: Common Fields for Cultivation. *Agriculture and Human Values* 19 (3): 217–224.

Desmarais, Annette Aurélie. 2007. *La Vía Campesina: Globalization and the Power of Peasants*. Halifax and London: Fernwood Publishing and Pluto Press.

Donald, B., and A. Blay-Palmer. 2006. The Urban Creative-Food Economy: Producing Food for the Urban Elite or Social Inclusion Opportunity. *Environment and Planning A* 38:1901–1920.

Duncan, C. 1996. *The Centrality of Agriculture: Between Humankind and the Rest of Nature*. Kingston: McGill-Queen's University Press.

DuPuis, M. 2002. *Nature's Perfect Food: How Milk Became America's Drink*. New York: New York University Press.

DuPuis, M., and D. Goodman. 2005. Should We Go "Home" to Eat? Towards a Reflexive Politics of Localism. *Journal of Rural Studies* 21:359–371.

Dyer, G., and J. C. Belausteguigoitia. 2006. Structural Adjustment, Market and Policy Failures: The Case of Maize. www.cbd.int/doc/case-studies/inc/cs-inc-iucn-06-en .pdf (accessed July 12, 2008).

Economist. 1996. Bigger than Baguettes 339 (7967), May 25, 1996: 73.

Edelman, M. 2008. Transnational Organizing in Agrarian Central America: Histories, Challenges, Prospects. *Journal of Agrarian Change* 8 (2–3): 229–257.

Elderidge, S. 2003. *Food Biotechnology: Current Issues and Perspectives*. New York: Nova Publishers.

Escobar, A. 1992. Culture, Economics and Politics in Latin American Social Movement Theory and Research. In *The Making of Social Movements in Latin America: Identity, Strategy and Democracy*. Boulder, CO: Westview Press.

Escobar, A. 1998. Whose Knowledge, Whose Nature? Biodiversity, Conservation, and the Political Ecology of Social Movements. *Journal of Political Ecology* 5:53–82.

Escobar, A. 1999. Biodiversity: A Perspective from Within. *Seedling* 16 (2): 24–31.

Escobar, A. 2001. Culture Sits in Places. Reflections on Globalism and Subaltern Strategies of Globalization. *Political Geography* 20:139–174.

Escobar, A. 2004. Beyond the Third World: Imperial Globality, Global Coloniality and Anti-globalisation Social Movements. *Third World Quarterly* 25 (1): 207–230.

Escobar Moreno, D. A. 2005. Valoracion de la Agrobiodiversidad: Una Aproximacion desde la Economia Ecologica. *Geografia Agricola* 35:7–22.

Escobar Moreno, D. 2007. El Valor Multifuncional del Agroecosistema Milpa en México. Paper presented at the Latin American Studies Association Meeting, Montreal, September 5–8.

Esteva, G. 1996. Hosting the Otherness of the Other: The Case of the Green Revolution. In *Decolonizing Knowledge: From Development to Dialogue*, ed. S. Marglin and F. Apfel-Marglin, 62–87. Clarendon: Oxford University Press.

Esteva, G. 2005. *Celebration of Zapatismo*. Oaxaca: Ediciones ¡Basta!

Esteva, G. 2007. The Other Campaign, APPO, and the Left. *Z Magazine*, March. www .zmag.org/zmag/viewArticle/13751 (accessed June 20, 2008).

Esteva, G., and C. Marielle, eds. 2003. *Sin Maiz no Hay Pais*. Mexico City: Museo Nacional de Culturas Populares.

Esteva, G., and M. S. Prakash. 1998. *Grasroots Postmodernism: Remaking the Soil of Cultures*. London: Zed Books.

Fairbairn, M. 2010. Framing Resistance: International Food Regimes and the Roots of Food Sovereignty. In *Food Sovereignty: Reconnecting Food, Nature and Community*, ed. H. Wittman, A. A. Desmarais, and N. Wiebe. Toronto: Fernwood Publishing.

Featherstone, D. 2007. Spatialities of Transnational Resistance to Globalization: The Maps of Grievance of the Inter-Continental Caravan. *Transactions of the Institute of British Geographers* 28 (4): 404–421.

Field, D. 2007. FoodShare Annual Meeting. Toronto: FoodShare.

Fitting, E. 2006. Importing Corn, Exporting Labor: The Neoliberal Corn Regime, GMOs, and the Erosion of Mexican Biodiversity. *Agriculture and Human Values* 23:15–26.

Fitting, E. 2007. Comments on the "Transgenic Maize Debates in Mexico" Panel. Paper presented at the Latin American Studies Association Meeting, Montreal, September 5–8.

FitzSimmons, M., and D. Goodman. 1998. Incorporating Nature: Environmental Narratives and the Reproduction of Food. In *Remaking Reality: Nature and the Millennium*, ed. B. Braun and N. Castree, 193–219. London: Routledge.

Fonte, M. 2006. Slow Food's Presidia: What Do Small Producers Do with Big Retailers? *Research in Rural Sociology and Development* 12:1–39.

Food and Agriculture Organization (FAO). 2003. Trade and Food Security: Conceptualizing the Linkages Expert Consultation. www.fao.org/DOCREP/005/Y4671E/y4671e03.htm#TopOfPage (accessed June 9, 2008).

Food Democracy. 2007. Say "Bye" to BYE: Monsanto GM Corn to Be Subsidized by USDA Federal Crop Insurance Program (FCIC). December 5. fooddemocracy.wordpress.com/2007/12/05/say-bye-to-bye-monsanto-gm-corn-to-be-subsidized-by-usda-federal-crop-insurance-program-fcic/ (accessed December 5, 2007).

FoodFirst. 2003. Food Sovereignty: Global Rallying Cry of Farmer Movements. *FoodFirst Backgrounder* 9(4). www.nyeleni.org/IMG/pdf/FoodSovRosset.pdf (accessed May 8, 2005).

Food Sovereignty: A Right for All. 2002. Rome NGO/CSO Forum for Food Sovereignty. www.nyeleni2007.org/spip.php?article125 (accessed September 4, 2007).

Forget, G., and J. Lebel. 2001. An Ecosystem Approach to Human Health. *International Journal of Occupational and Environmental Health* Supplement 7 (1): S3–S38.

Forsyth, T. 2003. *Critical Political Ecology: The Politics of Environmental Science.* London: Routledge.

Fowler, C., and P. Mooney. 1996. *Shattering: Food, Politics, and the Loss of Genetic Diversity.* Tucson: University of Arizona Press.

Freidburg, S. 2005. *French Beans and Food Scares: Culture and Commerce in an Anxious Age.* London: Oxford University Press.

Friedmann, H. 1980. Household Production and the National Economy: Concepts for the Analysis of Agrarian Formations. *Journal of Peasant Studies* 7:158–184.

Friedmann, H. 1982. The Political Economy of Food: The Rise and Fall of the Postwar International Food Order. *American Journal of Sociology* 88:248–286.

Friedmann, H. 1993. New Wines, New Bottles: Regulation of Capital on a World Scale. In *Production, Space, Identity: Political Economy Faces the 21st Century*, ed. J.

Jenson, R. Mahon, and M. Bienefeld. Toronto: Canadian Scholars Press. (Reprinted from Studies in Political Economy 36 [1991]. Greenwood, CT: Praeger.)

Friedmann, H. 2003. Eating in the Gardens of Gaia: Envisioning Polycultural Communities. In *Fighting for the Farm: Rural America Transformed*, ed. J. Adams, 252–273. Philadelphia: University of Pennsylvania Press.

Friedmann, H. 2004. Feeding the Empire: The Pathologies of Globalized Agriculture. In *The Empire Reloaded: Socialist Register 2005*, ed. C. Leys and L. Panitch. London: Merlin Press.

Friedmann, H. 2005a. From Colonialism to Green Capitalism: Social Movements and Emergence of Food Regimes. *New Directions in the Sociology of Global Development, Research in Rural Sociology and Development* 11:229–267.

Friedmann, H. 2005b. Modernity and the Hamburger: Cattle and Wheat in Ecological and Culinary Change. Paper presented at the colloquium in agrarian studies, Yale University, January 14. www.yale.edu/agrarianstudies/colloqpapers/Hamburger.pdf (accessed June 25, 2005).

Friedmann, H., and A. McNair. 2008. Whose Rules Rule? Contested Projects to Certify "Local Production for Distant Consumers." *Journal of Agrarian Change* 8 (2–3): 408–434.

Fussell, Betty. 1999. Translating Maize into Corn: The Transformation of America's Native Grain. *Social Research* 66 (1): 41–65.

Ganz, D. 2003. Barefoot Silviculture in Oaxaca, Mexico: The Adaptation of Standard Silvicultural Systems for Non-Timber Forest Products and the Integration of Indigenous Knowledge. www.fao.org/DOCREP/005/Y4496E/Y4496E25.htm (accessed October 23, 2006).

Gibson-Graham, J. K. 1996. *The End of Capitalism (As We Knew It): A Feminist Critique of Political Economy*. Oxford, UK, and Cambridge, MA: Blackwell.

Gibson-Graham, J. K. 2006. *A Postcapitalist Politics*. Minneapolis: University of Minnesota Press.

Gilbert, L., and C. Phillips. 2003. Practices of Urban Environmental Citizenships: Rights to the City and Rights to Nature in Toronto. *Citizenship Studies* 7 (3): 313–330.

Giuliani, A. 2007. *Developing Markets for Agrobiodiversity: Securing Livelihoods in Dryland Areas*. London: Earthscan.

Gliessman, S. 2006. *Agroecology: The Ecology of Sustainable Food Systems, Second Edition*. Boca Raton, FL: Taylor and Francis.

González, R. J. 2001. *Zapotec Science: Farming and Food in the Northern Sierra of Oaxaca*. Austin: University of Texas Press.

González, S. G. 2008. Denuncian Industriales de la Masa y la Tortilla Práctica Depredatoria en su Contra. *La Jornada*, May 18. www.jornada.unam.mx/2008/05/18/index.php?section=economia&article=020n1eco (accessed October 21, 2008).

Goodman, D. 2003. The Brave New Worlds of Agricultural Technoscience: Changing Perspectives, Recurrent Themes, and New Directions in Agro-Food Studies. In *Recreating the World: Genetic Engineering and Its Discontents*, ed. R. Schurman and D. Kelso. Berkeley: University of California Press.

Goodman, D., and E. M. DuPuis. 2002. Knowing Food and Growing Food: Beyond the Production-Consumption Debate in the Sociology of Agriculture. *Sociologia Ruralis* 42 (1): 5–22.

Goodman, D., and M. Redclift. 1991. *Refashioning Nature: Food, Ecology and Culture*. London: Routledge.

Goodman, D., and M. Watts. 1997. *Globalising Food: Agrarian Questions and Global Restructuring*. London: Routledge.

Greenpeace. 2011. The Fight Is Still Going on Over Mexican Maize. *Making Waves* (blog), March 2. www.greenpeace.org/international/en/news/Blogs/makingwaves/the-fight-is-still-going-on-over-mexican-maiz/blog/33535.

Gupta, A. 1998. *Postcolonial Developments: Agriculture in the Making of Modern India*. Raleigh, NC: Duke University Press.

Guthman. 2004. *Agrarian Dreams? The Paradox of Organic Farming in California*. Berkeley: University of California Press.

Hamilton, N. 2011. *Mexico: Political, Social and Economic Evolution*. London: Oxford University Press.

Haraway, D. 1997. *Modest_Witness@Second_Millennium.FemaleMan©Meets_Onco-Mouse: Feminism and Technoscience*. New York: Routledge.

Harvey, D. 1996. *Justice, Nature and the Geography of Difference*. Oxford, UK: Blackwell.

Harvey, D. 2000. *Spaces of Hope*. Berkeley: University of California Press.

Hassanein, N. 2003. Practicing Food Democracy: A Pragmatic Politics of Transformation. *Journal of Rural Studies* 19 (1): 77–86.

Henriques, G., and R. Patel. 2004. NAFTA, Corn, and Mexico's Agricultural Trade Liberalization. Americas Program Special Report. www.americaspolicy.org (accessed April 12, 2007).

Hinrichs, C. 2003. The Practice and Politics of Food System Localization. *Journal of Rural Studies* 19:33–45.

Hitchings, R., and V. Jones. 2004. Living with Plants and the Exploration of Botanical Encounter within Human Geographic Research Practice. *Ethics, Place and Environment* 7 (1–2): 3–18.

Hobbs, R. 2003. *Mark Lombardi: Global Networks*. New York: Independent Curators International.

Holloway, L., M. Kneafsey, L. Venn, R. Cox, E. Dowler, and H. Tuomainen. 2005. Possible Food Economies: Food Production-Consumption Arrangements and the Meaning of "Alternative." Working Paper 25. Cultures of Consumption Working Paper Series. Cultures of Consumption, and ESRC-AHRC Research Programme.

Holt-Giménez, E. 2006. Campesino a Campesino: Voices from Latin America's Farmer to Farmer Movement for Sustainable Agriculture. Oakland, CA: Food First Books.

hooks, b. 1994. *Teaching to Transgress: Education as the Practice of Freedom*. London: Routledge.

Jacobs, J. 1996. *Edge of Empire: Postcolonialism and the City*. London and New York: Blackwell.

Jarosz, L. 2000. Understanding Agrifood Networks as Social Relations. *Agriculture and Human Values* 17:279–283.

Johnston, J. 2003. Who Cares about the Commons? *Capitalism, Nature, Socialism* 14 (4): 1–41.

Johnston, J., and L. Baker. 2005. Eating Outside the Box: FoodShare's Good Food Box and the Challenge of Scale. *Agriculture and Human Values* 22 (3): 1–13.

La Jornada Michoacán. 2011. El Congreso, por Aprobar la Ley de Tomento y Protección del Maíz Criollo, Eduardo Ferrer. January.

Kapoor, I. 2008. *The Postcolonial Politics of Development*. New York: Routledge.

Keith, M., and S. Pile. 1993. Introduction Part 1: The Politics of Place. In *Place and the Politics of Identity*, ed. M. Keith and S. Pile, 1–21. London: Routledge.

Kingsolver, B., et al. 2007. *Animal, Vegetable, Miracle: A Year of Food Life*. Boston: HarperCollins.

Kloppenburg, J. 2004. *First the Seed: The Political Economy of Plant Biotechnology*. Madison: University of Wisconsin Press.

Kneen, B. 2011. Citizen, Subject, and Food Sovereignty. *Rams Horn*, no. 284 (October 5).

Koc, M., R. MacRae, J. Welsh, and L. J. A. Mougeot. 1999. *For Hunger Proof Cities: Sustainable Urban Food Systems*. Ottawa: IDRC.

Korves, R. 2007. U.S.-Mexican Corn Trade Becomes Free in 2008. http://www.truthabouttrade.org/2007/12/14/u-s-mexican-corn-trade-becomes-free-in-2008/ (accessed April 18, 2008).

LaDuke, W. 2002. *The Winona LaDuke Reader*. Osceola, WI: Voyageur Press.

LaDuke, W. 2005. *Recovering the Sacred: The Power of Naming and Claiming*. Cambridge, MA: South End Press.

Lang, T., and M. Heasman. 2004. *Food Wars: The Global Battle for Mouths, Minds, and Markets*. London: Earthscan.

Latour, B. 2005. *Reassembling the Social: An Introduction to Actor-Network Theory*. Oxford: Oxford University Press.

Lawson, L. 2005. *City Bountiful: A History of Urban-Garden Programs in America, 1890s to Present*. Berkeley: University of California Press.

Leyshon, A., and R. Lee. 2003. Introduction: Alternative Economic Geographies. In *Alternative Economic Spaces*, ed. A. Leyshon, R. Lee, and C. C. Williams, 1–27. London: Sage Publications.

Leyshon, A., R. Lee, and C. C. Williams, eds. 2003. *Alternative Economic Spaces*. London: Sage Publications.

Li, T. 2004. Environment, Indigeneity and Transnationalism. In *Liberation Ecologies*, ed. R. Peet and M. Watts, 339–370. London: Routledge.

Lind, D., and E. Barham. 2004. The Social Life of a Tortilla: Food, Cultural Politics, and Contested Commodification. *Agriculture and Human Values* 21:47–60.

Lipsitz, G. 2005. Introduction in *Mangos, Chiles and Truckers: The Business of Transnationalism*, by G. Alvarez. Minneapolis, MN: University of Minnesota Press.

Lotter, A. 2004. NAFTA Is Underlying Cause of GE Corn Contamination in Mexico. Organic Consumers Association. Retrieved from http://www.organicconsumers.org/ge/naftacorn090704.cfm.

Maffi, L. 2008. Cultural Vitality: The Concept of Biocultural Diversity. *Resurgence* 250 (September/October): 10–12.

Mann, C. 2004. *Diversity on the Farm*. Boston: Political Economy Research Institute. University of Massachusetts, Ford Foundation.

Mann, C. 2005. *1491: New Revelations of the Americas before Columbus*. New York: Knopf.

Mansfield, B. 2003. Fish, Factory Trawlers, and Imitation Crab: The Nature of Quality in the Seafood Industry. *Journal of Rural Studies* 19 (1): 9–21.

Martínez, J., and I. Jacaranda. 2010. Las demandas de las organizaciones purépechas y el movimiento indígena en Michoacán. Universidad de Ciencias y Artes de Chiapas. *Estudios Sociales y Humanísticos* 8 (1): 64–79.

Massey, D. 1994. Geographies of Responsibility. *Geografiska Annaler: Series B, Human Geography* 86 (1): 5–18.

Massieu Trigo, Y. 2007. Genetic Diversity of Maize in Mexico: Transgenic Maize Pertinence in the Center of Origin. Paper presented at the Latin American Studies Association Meeting, Montreal, September 5–8.

Maxwell, S. 1996. Food Security: A Post-modern Perspective. *Food Policy* 21 (2): 155–170.

Maxwell, S., and T. Frankenberger. 1992. *Household Food Security: Concepts, Indicators and Measurements: A Technical Review*. New York and Rome: UNICEF and IFAD.

Maxwell, S., and M. Smith. 1992. Household Food Security: A Conceptual Review. In *Household Food Security: Concepts, Indicators, and Measurements: A Technical Review*, ed. S. Maxwell and T. Frankenberger. New York and Rome: UNICEF and IFAD.

McAfee, K. 2004. Geographies of Risk and Difference in Crop Genetic Engineering. *Geographical Review* 41 (91): 80–105.

McAfee, K. 2006. Sustainability and Social Justice in the Global Food System: Linking Conservation, Agroecology, and Food Security. www.kmcafee.com/sustain_en.php (accessed July 2, 2007).

McMichael, P. 2008. Peasants Make Their Own History, but Not Just as They Please. *Journal of Agrarian Change* 8 (2–3): 205–228.

Mexico. 2011. E-Local (government of Mexico website). www.e-local.gob.mx/wb2/ELOCAL/EMM_michoacan.

Michoacán. 2011a. Cambio de Michoacán (website). www.cambiodemichoacan.com.mx/vernota.php?id=77924.

Michoacán. 2011b. Estado de Michoacán website. http://www.centrodeagronegocios.com/.

Michoacán Orgánico. 2008. Organizacion y Analisis de la Informacion: Fundacion Produce Michoacán. www.metrocert.com/files/parteii.pdf.

Miller, S. 2008. *Edible Action: Food Activism and Alternative Economies*. Toronto: Fernwood Publishing.

Moore, D. 1997. Remapping Resistance: "Ground for Struggle" and the Politics of Place. In *Geographies of Resistance*, ed. S. Pile and M. Keith, 87–106. London: Routledge.

Moore, D. 1999. The Crucible of Cultural Politics: Reworking Development in Zimbabwe's Eastern Highlands. *American Ethnologist* 26 (3): 654–689.

Morgan, K., T. Marsden, and J. Murdoch. 2006. *Worlds of Food: Place, Power and Provenance in the Food Chain*. London: Oxford University Press.

Nadal, A. 2000. Mexican Corn: Genetic Variability and Trade Liberalization. Programa Sobre Ciencia, Technologia y Desarrollo Documento de Trabajo No 1–06. Mexico: Procientec, El Colegio de México.

Nadal, A. 2006. Mexico's Corn-Producing Sector: A Commentary. *Agriculture and Human Values* 23:33–36.

National Research Council. 1989. *Lost Crops of the Incas: Little-Known Plants of the Andes with Promise for Worldwide Cultivation*. Washington, DC: National Academy Press.

Nazarea, V. 1998. *Cultural Memory and Biodiversity*. Tucson: University of Arizona Press.

Nazarea, V. 2005. *Heirloom Seeds and Their Keepers: Marginality and Memory in the Conservation of Biological Diversity*. Tucson: University of Arizona Press.

Ocampo, S. 2007. El Alza al Precio de la Tortilla Podría Detonar Estallidos Sociales: UNORCA. *La Jornada*, January 27. www.jornada.unam.mx/2007/01/27/index.php?section=politica&article=014n1pol (accessed January 30, 2007).

Organic Consumers Association. 2008. Asciende Producción Orgánica en Michoacán a Más de 500 MDP. Vía Orgánica, November 17. http://www.organicconsumers.org/ACO/articulos/article_15657.cfm (accessed September 16, 2012)

Paczka, R. O. 2003. La Diversidad del Maiz en Mexico. In *Sin Maiz no Hay Pais*, ed. G. Esteva and C. Marielle, 123–154. Mexico City: Museo Nacional de Culturas Populares.

PANOS Institute, World Wildlife Fund. 2002. *Voces de las Montañas: Testimonios Orales de la Sierra Norte de Oaxaca*. London: PANOS Institute.

Paulson, S., L. Gezon, and M. Watts. 2005. Politics, Ecologies, Genealogies. In *Political Ecology across Places, Scales, and Social Groups*, ed. S. Paulson, L. Gezon, and M. Watts, 17–39. Brunswick, NJ: Rutgers University Press.

Pechlaner, G., and G. Otero. 2008. The Third Food Regime: Neoliberal Globalism and Agricultural Biotechnology in North America. *Sociologia Ruralis* 48 (4): 351–371.

Peet, R., and M. Watts. 2004. *Liberation Ecologies*. 2nd ed. London: Routledge.

Peña, D. 1998. *Chicano Culture, Ecology, Politics: Subversive Kin*. Tucson: University of Arizona Press.

Perales, R. H., S. B. Brush, and C. O. Qualset. 2003. Landraces of Maize in Central Mexico: An Altitudinal Transect. *Economic Botany* 57 (1): 7–20.

Petrini, C. 2005. *Slow Food Nation: Why Our Food Should Be Good, Clean, and Fair*. New York: Rizzoli Ex Libris.

Pilcher, J. 1998. *Que Viven los Tamales: Food and the Making of Mexican Identity*. Albuquerque: University of New Mexico Press.

Pilcher, J. 2005. Industrial Tortillas and Folkloric Pepsi: The Nutritional Consequences of Hybrid Cuisines in Mexico. In *The Cultural Politics of Food and Eating: A Reader*, ed. J. Watson and M. Caldwell, 235-250. Oxford: Blackwell.

Pile, S., and M. Keith. 1997. *Geographies of Resistance*. London: Routledge.

Polanyi, K. 2001, 1944. *The Great Transformation: The Political and Economic Origins of Our Time*. Boston: Beacon Press.

Pollan, M. 2002. *The Botany of Desire: A Plant's Eye View of the World*. New York: Random House.

Pollan, M. 2006. *The Omnivore's Dilemma: A Natural History of Four Meals*. New York: Thorndike Press.

Posey, D. A., ed. 1999. *Cultural and Spiritual Values of Biodiversity*. London: United Nations Environmental Programme and Intermediate Technology Publications.

Preibisch, K. L., G. R. Herrejón, and S. L. Wiggins. 2002. Defending Food Security in a Free Market Economy: The Gendered Dimensions of Restructuring in Rural Mexico. *Human Organization* 61 (1): 68–79.

Pretty, J. 2002. *Agri-Culture: Reconnecting People, Land and Nature*. London: Earthscan Publishers.

Pulido, L. 1996. *Environmentalism and Economic Justice: Two Chicano Struggles in the Southwest*. Tucson: University of Arizona Press.

Quist, D., and I. H. Chapela. 2001. Transgenic DNA Introgressed into Traditional Maize Landraces in Mexico. *Nature* 414:541–543.

Reh, E. 1939. Notes on Mixtec Indian Land and Farming Practices. *Primitive Man* 12 (1): 1–11.

Ribeiro, S. 2004. The Day the Sun Dies: Contamination and Resistance in Mexico. *Seedling*, July. www.mindfully.org/GE/2004/Mexico-Contamination-Resistance 16aug04.htm (accessed September 18, 2004).

Riches, G. 1999. Advancing the Human Right to Food in Canada: Social Policy and the Politics of Hunger, Welfare, and Food Security. *Agriculture, Food and Human Values* 16:203–211.

Riedemann, C. S., and F.E. Haucuja. 2007. El Maiz amarillo cultivado bajo contrato en Mexico 2000–2005: Reflexiones sobre un programa de comercializacion de granos. Universidad de Granada: *Cuadernos Geograficos*: 107–132.

Rivera, J. G., C. A. G. Aldapa, M. A. C. Villegas, and M. E. R. Garcia. 2004. Proceedings from 1er Congreso Nacional de Nixtamalizacion: Del Maiz a la Tortilla, Queretaro, Qro, Mexico, October 17–21.

RMALC (Mexican Network against Free Trade). 2008. Declaración Política del VII Foro Mesoamericano de los Pueblos ¡Mesoamérica en Resistencia, No al Neocolonialismo del Libre Comercio! www.rmalc.org.mx/index.shtml (accessed May 12, 2008).

Rocha, C. 2001. Urban Food Security Policy: The Case of Belo Horizonte, Brazil. *Journal for the Study of Food and Society* 5 (1): 36–47.

Rocheleau, D. 2007a. Political Ecology in the Key of Policy: From Chains of Explanation to Webs of Relation. *Geoforum* 39 (2): 716–727.

Rocheleau, D. 2007b. Rooted Networks, Relational Webs and Powers of Connection: Rethinking Human and Political Ecologies. *Geoforum* 38 (3): 433–437.

Rocheleau, D., L. Ross, J. Morrobel, L. Malaret, R. Hernandez, and T. Kominiak. 2001. Complex Communities and Emergent Ecologies in the Regional Agroforest of Zambrana-Chacuey, Dominican Republic. *Cultural Geographies* 8 (4): 465–492.

Rocheleau, D., B. Thomas-Slayter, and E. Wangari, eds. 1996. *Feminist Political Ecology: Global Perspectives and Local Experience*. London: Routledge.

Royal Society et al. 2000. Genetically Modified Plants for Food Use and Human Health. Policy Document 4/02. London: The Royal Society.

Ruggles, M. 2006. Maize as Medium of Communication. Paper presented at the CCA conference, Toronto, York University, June 1–3.

Salvador, A. P., B. M. Morales, and M. del V. Sanchez, eds. 2004. III Encuentro Internacional: Integracion Regional, Globalizacion y Sector Agropecuario. Mexico, Universidad Autonoma Chapingo.

Shiva, V. 1993. *Monocultures of the Mind: Biodiversity, Biotechnology and Agriculture.* New Delhi: Zed Press.

Slater, D. 1997. Spatialities of Power and Postmodern Ethics: Rethinking Geopolitical Encounters. *Environment and Planning D: Society and Space* 15 (1): 55–72.

Slow Food. http://www.slowfood.com (accessed April 15, 2007).

Smith, A., and J. MacKinnon. 2007. *100 Mile Diet.* Toronto: Random House Canada.

Stassart, P., S. Whatmore, and H. Renting. 2003. *Environment and Planning A: Special Issue on Alternative Food Networks: European Perspectives* 35(3).

Stedile, J. P., and B. M. Fernandes. 2004. *Gens Sans Terre: La Trajectoire du MST et la Lutte pour la Terre au Brésil.* Brazil: Le Temps Des Cerises.

Stevenson, G. 1998. Warrior Work and Builder Work: Exploring Responses to Bipolar Enterprise Structures in Advanced Agrifood Systems. Paper presented at the joint meetings of the Agriculture, Food, and Human Values Society and the Society for the Study of Food and Society, San Francisco, CA.

Sundberg, J. 2004. Identities in the Making: Conservation, Gender and Race in the Maya Biosphere Reserve, Guatemala. *Gender, Place and Culture* 11 (1): 43–66.

Tansey, G., and T. Rajotte, eds. 2008. *The Future Control of Food: A Guide to International Negotiations and Rules on Intellectual Property, Biodiversity and Food Security.* Ottawa: Earthscan/IDRC.

Tarasuk, V., and J. M. Eakin. 2003. Charitable Food Assistance a Symbolic Gesture: An Ethnographic Study of Food Banks in Ontario. *Social Science and Medicine* 56:1505–1515.

Tarrow, S. 2005. *The New Transnational Activism.* Cambridge: Cambridge University Press.

Toledo, V. 2001. Biocultural Diversity and Local Power in Mexico. In *On Biocultural Diversity: Linking Language, Knowledge, and the Environment*, ed. L. Maffi, 472–488. Washington, DC: Smithsonian Press.

Torres, G. 1997. *The Force of Irony: Power in the Everyday Life of Tomato Workers.* New York: Berg Publishers.

Tregear, A. 2003. From Stilton to Vimto: Using Food History to Re-think Typical Products in Rural Development. *Sociologia Ruralis* 43 (2): 91–107.

Trejo-Gonzalez, A., A. Feria-Morales, and C. Wild-Altamirano. 1982. The Role of Lime in the Alkaline Treatment of Corn for Tortilla Preparation. *Advances in Chemistry* 198 (9): 245–263.

Tsing, A. 2005. *Friction: An Ethnography of Global Connection.* Princeton, NJ: Princeton University Press.

Velázquez, R. B. 2005. Nuestro Maíz. *La Joranda*, January 17. www.jornada.unam.mx/2005/01/17/006n1sec.html (accessed October 21, 2008).

Verschoor, G. 2007. Framing the Controversy about GM Maize. In *Maize and Biosecurity in Mexico*, ed. E. Antal, L. Baker, and G. Verschoor, 33–54. Cuadernos de CEDLA, 22. Amsterdam: Centre for Latin American Research and Documentation.

Via Campesina. 2008. Background: La Via Campesina International Conference on Pesant Rights. http://viacampesina.org/en/index.php/main-issues-mainmenu-27/human-rights-mainmenu-40/538-background-la-via-campesina-international-conference-on-pesant-rights (accessed July 5, 2008).

Villagomez, M. D. 1999. Grassroots Responses to Globalization: Mexican Rural and Urban Woman's Collective Alternatives. In *Women Working the NAFTA Food Chain: Women, Food and Globalization*, ed. D. Barnt, 209–220. Toronto: Second Story Press.

Vizcarra Bordi, I. 2006. The "Authentic" Taco and Peasant Women: Nostalgic Consumption in the Era of Globalization. *Culture & Agriculture* 28 (2): 97–107.

Walker, P. A. 2007. Political Ecology: Where Is the Politics? *Progress in Human Geography* 31 (3): 363–369.

Waltner-Toews, D. 2004. *Ecosystem Sustainability and Health: A Practical Approach.* Cambridge: Cambridge University Press.

Warman, A. 2003. *Corn and Capitalism: How a Botanical Bastard Grew to Global Dominance.* Chapel Hill: University of North Carolina Press.

Watson, J., and M. Caldwell. 2005. *The Cultural Politics of Food and Eating: A Reader.* Oxford: Blackwell.

Watts, M., and D. Goodman. 1997. Agrarian Questions: Global Appetite, Local Metabolism: Nature, Culture, and Industry in Fin-de-Siecle Agrifood Systems. In *Globalising Food: Agrarian Questions and Global Restructuring*, ed. D. Goodman and M. Watts. London: Routledge.

Watts, M., B. Ilbery, and D. Maye. 2005. Making Reconnections in Agro-food Geography: Alternative Systems of Food Provision. *Progress in Human Geography* 29 (1): 22–40.

Weedon, C. 1997. *Feminist Practice and Poststructuralist Theory*. 2nd ed. Oxford: Blackwell.

Wekerle, G. R. 2004. Food Justice Movements: Policy, Planning, and Networks. *Journal of Planning Education and Research* 23:378–386.

Welsh, J., and R. MacRae. 1998. Food Citizenship and Community Food Security: Lessons from Toronto, Canada. *Canadian Journal of Development Studies* 19:237–255.

Whatmore, S. 2002. *Hybrid Geographies: Natures, Cultures, Spaces.* London: Sage.

Whatmore, S. 2004. Humanism's Excess: Some Thoughts on the "Post-human/ist" Agenda. *Environment and Planning A* 36: 1341–1363.

Whatmore, S., and P. Stassart. 2003. Metabolising Risk: Food Scares and the Un/remaking of Belgian Beef. *Environment and Planning A* 35 (3): 449–462.

Whatmore, S., and L. Thorne. 1997. Nourishing Networks. In *Globalising Food: Agrarian Questions and Global Restructuring*, ed. D. Goodman and M. Watts, 287–304. New York: Routledge.

Williams, R. 1973. *The Country and the City.* New York: Oxford University Press.

Winson, T. 1994. *The Intimate Commodity: Food and the Development of the Agro-Industrial Complex in Canada.* Toronto: University of Toronto Press.

Winter, M. 2003. Embeddedness, the New Food Economy and Defensive Localism. *Journal of Rural Studies* 19 (1): 23–32.

Winter, M. 2005. Geographies of Food: Agro-food Geographies—Food, Nature, Farmers and Agency. *Progress in Human Geography* 29 (5): 609–617.

Wise, T. 2004. The Paradox of Agricultural Subsidies: Measurement Issues, Agricultural Dumping, and Policy Reform. Global Development and Environment Institute Working Paper 04–02, Tufts University.

Wise, T. 2007. Policy Space for Mexican Maize: Protecting Agro-biodiversity by Promoting Rural Livelihoods. Global Development and Environment Institute Working Paper 07–01, Tufts University.

Wittman, H., A. A. Desmarais, and N. Wiebe, eds. 2010. *Food Sovereignty: Reconnecting Food, Nature and Community*. Toronto: Fernwood Publishing.

Woods, M. 2007. Engaging the Global Countryside: Globalization, Hybridity and the Reconstitution of Rural Place. *Progress in Human Geography* 31 (4): 485–507.

Yin, R. 1994. *Case Study Research Design and Methods, Applied Social Research Methods Series*. London: Sage.

Yunez-Naude, A. 2002. Lessons from NAFTA: The Case of Mexico's Agricultural Sector, Final Report to the World Bank. ctrc.sice.oas.org/geograph/north/yunez.pdf (accessed May 25, 2007).

Zimmerer, K. 2000. The Reworking of Conservation Geographies: Nonequilibrium Landscapes and Nature-Society Hybrids. *Annals of the Association of American Geographers* 90 (2): 356–369.

List of Interviews

Aguirre Alonso, Jorge. Crop Improvement Coordinator, Nuestro Maíz. February 25–27, 2006.

Antal, Edit. Researcher, Centre for Research on North America, National Autonomous University of Mexico. September 14, 2006.

Appendini, Kirsten. Professor, El Colegio de Mexico. January 24, 2006.

Aragón Cuevas, Flavio. Engineer. National Institute for Forestry, Agricultural and Livestock Research (INIFAP), Oaxaca Office. February 8. 2006.

Bautista Bautista, Patricia. Tortilla maker, Itanoní Tortillería. February 27, 2006.

Blas, Homero. Director, Organic Crop Improvement Association (OCIA). March 3, 2006.

Castellano Ortega, Luisa. Farmer, El Trapiche, Zimatlán, Oaxaca. February 15, 2006.

Castro, Humberto. Professor, Universidad de Chapingo. February 28, 2006.

Castro Moreno, Sandra (pseudonym). Tortilla maker, Nuestro Maiz Tortilleria. February 26, 2006.

Díaz Castellano, Hector. Farmer, El Trapiche, Zimatlán, Oaxaca. February 15, 2006.

Escobar Moreno, Darío. Researcher, independent. February 23, 2006.

Esteva, Gustavo. Professor, Universidad de la Tierra. February 17, 2006; June 24, 2008.

Fernandez Fernandez, Sabilla. Cook, Itanoní Tortillería. February 27, 2006.

Fernández Orantes, Lea Gabriela. Owner, Itanoní Tortillería. February 16, 2006.

Friedmann, Harriet. Professor, University of Toronto. April 24, 2008.
García Rodriguez, Judith Leticia. Researcher, World Wildlife Fund, Voces de las Montañas: Testimonios Orales de la Sierra Norte de Oaxaca Project. January 14, 2006.
Giaonetto, Fulvio. Farmer, Coyote Rojo. May 9, 2008.
Gómez Herrera, Roberto (pseudonym). Merchant/Farmer, San Antonio Atotonilco. February 25, 2006.
González, Aldo. President, Zapotec Union of Organization of the Sierra Juárez of Oaxaca (UNOSJO). March 2, 2006.
González, Edith Ruiz. Coordinator, Zapotec Union of Organization of the Sierra Juárez of Oaxaca (UNOSJO). March 2, 2006.
Herrera, Alejandro (pseudonym). Staff, Itanoní Tortillería. March 3, 2006.
Jimenez, Arturo (pseudonym). Staff, Nuestro Maiz Nixtamal Plant. February 27, 2006.
León Santos, Jesús. Director, Centro de Desarrollo Integral Campesino de la Mixteca (CEDICAM). March 6, 2006.
Paczka, Rafael. Professor, Universidad de Chapingo. January 25, 2006.
Perlas Salvador, Arturo. Professor, Universidad de Chapingo. January 25, 2006.
Poceros, J. Armando Joffre. Director, Nuestro Maíz. February 26, 2006.
Ramero, Luís (pseudonym). Farmer/Member, San Antonio Atotonilco Producer Group. February 25, 2006.
Ramírez Leyva, Amado. Owner, Itanoní Tortillería. February 10, 2006; February 28, 2006; May 8, 2008.
Reyes, "Chucho" (pseudonym). Farmer, Tooxí, Nochixtlan, Mexteca Alta, Oaxaca. March 6, 2006.
Ribeiro, Silvia. Program Coordinator, ETC Group. January 13, 2005.
Salinas Moreno, Yolanda. National Institute for Forestry, Agricultural and Livestock Research (INIFAP), Chapingo. February 24, 2006.
Zarrate, David. Journalist, independent. January 24, 2006.

Index

About the Author

Lauren Baker lives in Toronto, Canada, with her husband and two sons. She has a passion for healthy soil, vibrant communities, social justice, and corn tortillas. Lauren is the coordinator of the Toronto Food Policy Council at Toronto Public Health and teaches at the University of Toronto. Her past work includes being the founding director of Sustain Ontario—the Alliance for Healthy Food and Farming. In 1997, Lauren cofounded Toronto's first certified organic urban farm, which led to the creation of FoodShare Toronto's urban agriculture program. Lauren is a research associate with Ryerson University's Centre for Studies in Food Security and has a PhD in Environmental Studies from York University.